You Can Prevent Global Warming
(and save money!)

51 EASY WAYS

Jeffrey Langholz, Ph.D., and Kelly Turner

**Andrews McMeel
Publishing, LLC**

Kansas City

ISBN-13: 978-0-7407-7716-5
ISBN-10: 0-7407-7716-5

Library of Congress Cataloging-in-Publication Data
    Langholz, Jeff A. (Jeff Alan)
        You can prevent global warming (and save money!) : 51 easy ways / Jeffrey Langholz and Kelly Turner.
            p. cm.
        Includes bibliographical references.
        ISBN-13: 978-0-7407-7716-5
        ISBN-10: 0-7407-7716-5
        1. Environmental protection—Citizen participation. 2. Global warming—Prevention—Citizen participation. 3. Consumer education. I. Turner, Kelly, 1979- II. Title.
        TD171.7.L36 2008
        363.738'74525—dc22   2002044874

08 09 10 11 12 RR2 10 9 8 7 6 5 4 3 2 1

ATTENTION: SCHOOLS AND BUSINESSES
Andrews McMeel books are available at quantity discounts with bulk purchase for educational, business, or sales promotional use. For information, please write to: Special Sales Department, Andrews McMeel Publishing, LLC, 4520 Main Street, Kansas City, Missouri 64111.

This book is printed with soy-based ink on recycled paper with Forest Stewardship Council certification. The cover of this book is printed on recycled stock and is recyclable.

Mixed Sources
Product group from well-managed forests, controlled sources and recycled wood or fiber
www.fsc.org Cert no. SCS-COC-00648
© 1996 Forest Stewardship Council

FSC

A portion of the authors' proceeds from the sales of this book will be used to support educational programs in climate change.

To the members of the next generation,
who must live with the decisions we make today

Contents

Reduce, Reuse, Recycle, and More

Expand Your Impact

Introduction

The Bad News

It's a fact: The Earth is warming. For the first time since the dawn of human civilization, the Earth's temperature is rising. Scientists from around the world agree that the burning of fossil fuels contributes significantly to this warming, and they predict that the temperature will continue to rise as long as we keep burning these fuels. That's because when fossil fuels are burned, they release "greenhouse gases," which have a special ability to trap the sun's heat inside our atmosphere. The more fossil fuels we burn, the more greenhouse gases there are to trap the sun's heat—which means a warmer planet for us all.

A warmer planet might not sound all that bad to you, especially if you're from North Dakota. The effects of global warming won't stop there, however. A rise in the Earth's temperature will cause increasingly severe weather, including more sudden temperature swings, droughts, floods, heat waves, wildfires, and thunderstorms. As a result, our food and water supplies will be threatened and more people will die from events like heat waves and floods. Global warming will also cause the world's glaciers to melt, making sea levels rise and flooding low-lying areas such as the Mississippi Delta. A warmer Earth will endanger thousands of plant and animal species that cannot migrate to cooler regions. Tropical regions will expand, allowing insects that carry diseases such as malaria, dengue fever, and the West Nile virus to spread to places like

Florida, Georgia, or even New York. Scientists cannot predict exactly where or to what magnitude the consequences of global warming will occur, but they do know that hints of these effects are already happening.

As far as we know, humans are the most intelligent species ever to have walked planet Earth. However, our wonderful advances in civilization—transportation, heating, air-conditioning, and electricity—could very well contribute to our demise. The Earth's temperature has risen and fallen many times over the last four billion years. For example, many scientists believe that the dinosaurs became extinct after a giant asteroid hit the Earth and dramatically altered its climate. It is not an asteroid that threatens our extinction now, however. This time it is we humans—the *inhabitants* of Earth—who are altering the planet's climate, and we are doing so without fully understanding the consequences.

The Good News

There are easy ways you can prevent global warming and also save over $2,000 a year. Global warming is a colossal problem that is going to require the entire world's cooperation, but you don't have to feel helpless. Instead, you can help the planet *and* your wallet by following these 51 tips.

The Goal

We wrote this book for three reasons. First, we wanted to empower concerned citizens—to show them easy things they can do locally to help tackle an important global issue. Second, we wanted to put money back in their wallets—an average of $2,000 per household. This is especially important during difficult economic times. Finally, we wanted to show that economic growth and environmental protection can go hand in hand. In March 2001, the United States formally withdrew support for the international global warming treaty called the Kyoto Protocol (see Appendix B for more info). An important concern was that trying to reduce our nation's greenhouse gas emissions might hurt our economy. This book addresses that concern head on, showing clearly how we can prevent global warming *and* improve individual households' economic outlook. The overall goal is this: If enough people (35 percent of the U.S. population to be exact) follow the suggestions in this book, we can reduce our nation's emissions to the level the Kyoto Protocol targeted while improving individuals' financial pictures. As governments worldwide continue debating how to deal with global warming, individuals can choose to forge ahead now, helping both the planet and themselves. This book shows the way.

The Finish Line

Reading this book is supposed to be an enlightening experience, not a guilt trip. If you start to feel overwhelmed by all the things we're asking you to do, take a break and realize that you've made a difference simply by picking up this book and reading it. Then, flip to the end of the book, where we've put together a suggested order for following the 51 tips. Most of all, enjoy!

Notes on the Text

What We Mean by *Energy*

We burn the three fossil fuels (coal, petroleum, and natural gas) to provide us with energy: energy to power our cars, heat our homes, and create electricity. All the tips in this book focus on conserving energy, because when you conserve energy (in the form of gasoline, natural gas, or electricity), you reduce the demand for fossil fuels. When we reduce the demand for fossil fuels, less fossil fuel will be burned, and greenhouse gas emissions will decline. So, by telling you how to conserve "energy," we're telling you how to conserve fossil fuels.

What We Mean by *Carbon Dioxide* (CO_2)

There are 11 main greenhouse gases. In this book, we focus almost solely on carbon dioxide (CO_2), although we do mention ways to reduce two other greenhouse gases, methane and chlorofluorocarbons (CFC's). We focus mainly on CO_2 because it accounts for 52.5 percent of the world's global warming problem—more than all the other 10 greenhouse gases combined. In terms of global warming, CO_2 is measured in pounds or tons (one ton = 2,000 pounds). The following comparisons should help you put pounds of CO_2 into perspective:

- A barrel of oil that's burned emits 937 pounds of CO_2.

- A typical passenger car emits 12,000 pounds of CO_2 a year.

- A typical 500-megawatt coal-burning power plant emits three billion pounds of CO_2 a year.

What We Mean by *Kilowatt-Hour*

Throughout this book we will be talking about various sources of energy, such as natural gas, gasoline, and electricity. Natural gas is measured in therms, and gasoline is measured in gallons. Electricity, however, gets a bit tricky. You're probably familiar with the term *watt*. A watt is the measure of how quickly electricity is used at any given moment. For example, a 100-watt lightbulb will use 100 watts of electricity at any given moment. In order to measure how much electricity is used over a *period of time*, however, you need to add in a measure of time, such as the hour. A *watt-hour* (Wh) is the measure of how much electricity something (such as a lightbulb) uses in one hour. For example, a 100-watt lightbulb, which uses 100 watts of electricity at any given moment, will use 100 watt-hours of electricity in 1 hour or 200 watt-hours of electricity in 2 hours (or 150 watt-hours in 1.5 hours).

A watt is a very small unit of measure, though. That's why, in this book, we will be using kilowatts (1 kW = 1,000 watts) and kilowatt-hours (1 kWh = 1,000 watt-hours). Kilowatt-hours are what you see on your electricity bill. To find out how much your utility company charges you per kilowatt-hour, simply divide your monthly bill by the number of kilowatt-hours you used that month. The national average retail price of electricity in 2000 (which is what we used for all this book's calculations) was 8.16 cents per kilowatt-hour.

Climate Change Versus Global Warming

Global warming was the name first used to define the phenomenon of the Earth warming as a result of excess amounts of greenhouse gases in the atmosphere. Recently scientists have begun calling it climate change instead, since climate change more accurately describes what's happening: It's not just that the Earth is warming; rather the whole climate is being affected (more floods, droughts, heat waves, et cetera). In other words, global warming and climate change refer to the same thing.

The Companies and Organizations We Mention

We independently researched all the tips in this book. As a result, we recommend only brands, products, companies, organizations, and Internet sites that we personally believe are worthy of mention. We did not receive any type of compensation for referencing any product or company in this book.

Home Electricity and Hot Water

Light Up Your Life

Compact fluorescent lightbulbs last for up to 10 years and use 75 percent less energy than regular incandescent bulbs.

Overview. For some reason, Americans are being swindled. They are still using the same lightbulb Thomas Edison invented when they could be using one that lasts 10 times longer, uses one-fourth of the energy, and produces more light per watt. Why is this? No one is exactly sure—perhaps the compact fluorescent companies don't have big enough advertising budgets. Whatever the reason, now that you know, there's no excuse not to run to your hardware store today and buy these superefficient bulbs. Although compact fluorescent bulbs are initially more expensive than incandescent ones (around $15 each), they will quickly pay for themselves and start saving you money thanks to their long life.

What You Should Know

- Compact fluorescent bulbs are not the same as the fluorescents you see in schools and hospitals—you know, those stark, glaring tubes that buzz? *Compact* fluorescents are bulb shaped and give off warm light just like regular incandescent bulbs.

- A *lumen* measures the amount of light a lightbulb gives off, and a watt measures the amount of electricity a lightbulb uses. Compact fluorescents are so efficient because they give off the same amount of lumens, but use only one-fourth the watts that incandescent bulbs do. For example, a 25-watt compact fluorescent bulb has the same light output as a 100-watt incandescent bulb.

- Incandescent bulbs are inefficient because they give off 90 percent heat and only 10 percent light. Halogen lamps are only slightly more efficient, but they're also major fire hazards, reaching temperatures of up to 1,000°F. Your best option, in terms of efficiency *and* safety, is the compact fluorescent.

- Older versions of compact fluorescent (CFL) bulbs didn't always fit into incandescent light sockets. However, the recently invented subcompact fluorescents fit into almost *any* incandescent socket.

- *Climate Results:* If every household in the United States replaced its next burned-out lightbulb with a compact fluorescent, we would prevent more than 13 *billion pounds* of carbon dioxide from being emitted—that's equivalent to taking 1.2 million cars off the road for an entire year.

- *Money Matters:* Installing motion sensors on outdoor lights saves the average household $35 a year in electricity costs.

Easy Ways You Can Help

- *Turn off your lights* when you leave a room. Contrary to popular belief, it does *not* take more energy to turn a light off and then on again than it does just to leave it on (this goes for both incandescent and compact fluorescent bulbs).

- *Install motion sensors,* daylight sensors, or simple timers on your outdoor lights. You may need professional assistance to do this, but your annual energy savings will more than make up for the cost of installation.

- *Install dimmer switches* on indoor lights that don't always need to be at full brightness. A bulb dimmed by 50 percent will use half as much electricity, and dimming also extends the life of the bulb. You can pick up dimmer switches at your local hardware store and install them yourself.

- *Unique ways to make things brighter:* Dusting lightbulbs allows you to get away with using lower-wattage bulbs while having the room feel equally bright, as does painting your walls a lighter color, such as off-white or light blue. Lighter walls reflect up to 80 percent more light than dark walls do. Last, don't forget to take advantage of the brightest light in our universe: the sun! Open those curtains during the day, turn off your lights, and bask in the Earth's only natural source of light.

- *Switch to compact fluorescent bulbs!* This is the best thing you can do to reduce carbon dioxide emissions *and* your monthly energy bill. First, call your electric utility company to see if they offer rebates or incentives for buying compact fluorescents. Head to your hardware store, where you can get compact fluorescent bulbs, sconces, table lamps, or torchères (for your halogen floor lamps). Then, start replacing your incandescent bulbs, especially the ones that are on for more than two hours a day, and definitely those halogens! Compact fluorescents are also ideal for outdoor lighting since you'll only have to replace them once every 7 to 10 years.

Make sure you buy compact fluorescents with electronic, solid-state ballasts, which means that they turn on instantly and won't flicker or buzz. If you want to be able to dim your compact fluorescents, make sure you buy ones that are marked "dimmable." If you live in a colder climate, be sure to ask for outdoor compact fluorescents with cold-weather ballasts, so they'll function below 32°F. Last, don't forget to recycle the incandescent lightbulbs you are replacing (see Tip 47). Better yet, donate them to a local shelter or low-income housing development.

More Savings, Less Carbon Dioxide

Annual amount of money **saved** as a result of replacing 4 incandescent bulbs with CFL's, taking into account the initial cost of the CFL's:	Annual amount of CO_2 **not** *emitted* as a result of replacing 4 incandescent bulbs with CFL's:	*Total* amount of *money* **saved** over the life of the CFL's as a result of replacing 4 incandescent bulbs with CFL's, taking into account the initial cost of the CFL's:
$30	**718 lbs.**	**$205**

Assumptions: Four 100-watt incandescents that are used 4 hours per day, 365 days a year, are replaced by four 25-watt compact fluorescents; one CFL costs $14 and lasts 10,000 hours; one incandescent bulb costs $0.50 and lasts 1,000 hours.

Search For More Info

- www.compactoffer.com/product_listing.cfm?pc_id=xcel—Most hardware stores now carry compact fluorescents, but if you'd rather shop from home, you can buy them on-line at this web site.

- www.easy2diy.com/tutorials/diy0156/index.asp—Here is an illustrated do-it-yourself guide on installing a dimmer switch.

- www.ase.org/programs/torchiere.htm—Check out this web site to learn more about compact fluorescent halogen lamps and how they came about.

Rein in the Fridge

A refrigerator costs the average American household $120 a year in electricity.

Overview. You may think you're the pig when it comes to the refrigerator, but little did you know that your fridge is the biggest hog of your entire house—in terms of energy. This beast of an appliance sucks up 13 percent of your electricity each month! Fortunately, there are lots of little things you can do to rein in this energy animal, thereby saving money *and* carbon dioxide.

What You Should Know

- Keeping two smaller refrigerators is more expensive and less energy-efficient than keeping one large refrigerator.

- In 1995 the U.S. government was wise enough to realize that the chlorofluorocarbons in our refrigerators, air conditioners, and aerosol cans were creating a damaging hole in the Earth's ozone layer. Unfortunately, they decided to replace these chlorofluorocarbons with hydrochlorofluorocarbons, which are only *slightly* less damaging to the ozone layer and 1,800 times *more* damaging in terms of global warming!

- Brand-new, energy-efficient refrigerators use half as much energy as most 10-year-old fridges, which means they can reduce your refrigerator's electric bill by at least $60 a year—for the next 18 years.

- *Climate Results:* If every American household turned up their refrigerator temperature by 1°F, we would prevent almost three *million tons* of carbon dioxide from entering the atmosphere—every year.

- *Money Matters:* That old spare refrigerator you use in the basement or garage could be costing you more than $150 a year.

Easy Ways You Can Help

- *Choose a cool spot.* Don't make your fridge work harder than it needs to by locating it in direct sunlight or next to a heat source such as a dishwasher, stove, oven, or heating vent.

- *Don't get too chilly*. Many Americans' refrigerators are set colder than they need to be. Your fridge's thermostat dial should be set to 37 to 40°F, and your freezer's to zero to 5°F. Your food will still be *plenty* cold and safe from bacteria growth, plus you'll save a lot of electricity.

- *Turn on the "energy saver" switch.* If your refrigerator has an energy saver switch (often near the thermostat), turn it up as high as you can without having condensation form on the outside of your fridge. Also, if your fridge has a butter conditioner—a little heater that keeps your butter soft but also uses a lot of electricity—switch it to the hard setting. Better yet, turn it off altogether and just take the butter out a few minutes before you need it.

- **Junk that extra fridge!** If you're keeping an inefficient spare refrigerator or freezer in your basement just to cool a few beverages, seriously consider recycling it. A 10-year-old fridge could be costing you hundreds of dollars and emitting nearly one ton of carbon dioxide a year. Furthermore, if that extra fridge is in your garage, the coils will quickly collect dust and summer heat will damage the door seals—both of which cause your fridge to work harder. To save money and energy, either upgrade to a more efficient model or simply retire (and recycle!) the old clunker.

- **Upgrade and save.** You'll be shocked by how much money and energy you can save by upgrading to a new, energy-efficient fridge—even if yours *isn't* on its last legs. See Tip 14.

- **Demand the Greenfreeze.** Unhappy with the government's decision in 1995 to replace the chlorofluorocarbons in our refrigerators with hydrochlorofluorocarbons, the environmental group Greenpeace went on a mission to invent a completely environmentally safe refrigerator. They succeeded, and the Greenfreeze refrigerator is now the most popular fridge in Europe. It's not being sold in the United States, however, because our manufacturers don't feel like making it.

That's right, even though these fridges would be *cheaper* to manufacture in the long run, the companies can't be bothered to change their ways simply to help the environment. But *you* should be bothered—not only for global warming but also for money. The Greenfreeze is up to 38 percent more energy-efficient than an identical fridge that uses HCFC's, which means huge savings for you. Write or e-mail your congressional representative today (see Tip 51) asking that the government mandate an immediate phase-in of Greenfreeze technology and a phase-out of HCFC's.

- *Make sure the door seals are airtight.* If chilled air can escape through your fridge's door, it will have to work that much harder to replace it. Test the rubber seal (gasket) around the door every six months by closing the door on a piece of paper. You should feel some resistance as you pull the paper out. If you don't, you may need to have the seals replaced. Replacement can be quite expensive, however, so it may make more sense to upgrade to a new refrigerator (see Tip 14).

- *Clean those coils.* Your fridge stays cold because it has something called a condenser coil located on the bottom or back that removes heat from the inside. When the coil becomes dusty and dirty, it doesn't function as well, and therefore your fridge has to use more energy. Unless you have a "no-clean condenser" model (check your owner's manual), you should gently vacuum or brush off the coils twice a year.

Your fridge's efficiency will improve by up to 30 percent! Also, make sure that there are at least two inches of space around your entire fridge so that air can flow freely and ventilate the hot coils.

- **Keep the doors closed** as much as possible to hold the cold air inside. Organize and label your fridge's contents so you don't have to spend five minutes looking for the pickles!

- **Fill it up.** Keep your fridge filled but not overcrowded. When the door is opened, a full fridge will keep in cold air better than a partially filled fridge will. To fill up an empty fridge, put in extra pitchers of water. Don't go too crazy, though—an overcrowded fridge will prevent the cold air from circulating properly.

All the tips mentioned here apply to freezers too!

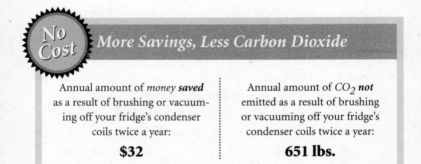

No Cost

More Savings, Less Carbon Dioxide

Annual amount of *money* **saved** as a result of brushing or vacuuming off your fridge's condenser coils twice a year:

$32

Annual amount of CO_2 **not** emitted as a result of brushing or vacuuming off your fridge's condenser coils twice a year:

651 lbs.

Assumptions: Based on a typical fridge that uses 1,323 kilowatt-hours per year at 8.16 cents per kilowatt-hour; cleaning condenser coils every six months improves efficiency by up to 30 percent.

Search for More Info

- www.pstvnrg.com/xina/rcc3.html—This web site has photos that show you exactly how to clean your refrigerator's coils.

- www.greenpeace.org/~climate/greenfreeze—Read all about Greenpeace's Greenfreeze refrigerator, and how U.S. chemical companies simply don't feel like manufacturing it.

- www.energystar.gov/products/refrigerators—Go here to find out how much the newest energy-efficient refrigerators and freezers can save you.

Don't Be a Washout

Eighty percent of the electricity a washing machine uses goes toward heating the water.

Overview. Still stuck on the idea that scalding water is what gets your clothes clean? Well, it does, but so does modern detergent. Back in the days of washboards, hot water was necessary to get rid of germs, but today Tide works just as well, if not better. Washing with cold water is just one of the ways you can improve the energy efficiency of your washing machine, and there is even more room for improvement with your dryer. The bad news is that you still have to do your laundry. The good news is that you can do it more cheaply and efficiently.

What You Should Know

- The average American household does nearly 400 loads of laundry per year, with machines that use 40 to 60 gallons of water per load. That's the equivalent of 650 showers a year.

- A dryer alone costs the average household about $90 a year in electricity.

- Using cold water to wash your clothes instead of hot water can cut your energy use—and your washing costs—in half!

- *Climate Results:* This year, if every American household hung just one load of laundry out on a clothesline to dry instead of using a dryer, they would prevent 250,000 *tons* of carbon dioxide from being emitted into the atmosphere.

- *Money Matters:* Using the moisture-sensor setting on your dryer can reduce your drying costs by 15 percent.

Easy Ways You Can Help

Washing Machine

- *Wash clothes in warm or cold water instead of hot.* Your clothes will come out just as clean, and you will cut your energy usage by 50 percent. Be sure to use a cold-water detergent (as most detergents are these days). Try pre-treating oily stains with a stain stick. For really dirty clothes, selecting a warm presoak cycle followed by cold wash and rinse cycles is still more energy-efficient than using a hot wash cycle.

- *Always* rinse *clothes in cold water.* Rinsing your clothes in warm water won't make them any cleaner—it's the wash cycle that does that. But rinsing in cold water *will* make your energy bill lower.

- *Don't use too much detergent.* Too many suds will make your washing machine work harder than it should to rinse out the clothes, and therefore use more energy. Be sure to use only the amount specified on the label.

- *Always wash a full load,* or select the "small load" or "low water level" setting (if your machine has one). Washing one large load is usually more efficient than washing two smaller loads.

- *Check to see if your washer has a horizontal axis.* If it doesn't, you should seriously consider upgrading. Horizontal-axis (H-axis) washing machines use 30 to 60 percent less water than standard vertical-axis machines and 50 to 70 percent less energy (this includes energy saved from drying time). See Tip 14 for more information.

Dryer

- *Line-dry clothes whenever you can.* Line drying is absolutely free and uses no energy (except your own). Your clothes will smell like they've been dried in the sunshine. If it's raining or below freezing, use an indoor drying rack or line.

- *Don't overdry your clothes.* Overdrying wastes energy and money, and wears down fabric. Select your dryer's automatic moisture-sensor setting or experiment with the timer to see how long it takes to dry a typical load. The moisture-sensor setting uses a computer chip to stop the dryer precisely when your clothes are dry, and it can reduce your energy use by 15 percent.

- *Always dry a full load,* but make sure there's enough room for the clothes to tumble. In the summer, don't run your dryer during the hottest part of the day. It will have to work harder and therefore use more energy.

- **Clean the lint trap before every load.** Cleaning the lint trap improves air circulation, which makes it easier for your dryer to dry your clothes. Also, check the exhaust hose every few months to make sure it's not blocked or full of lint.

- **Select the cool-down feature,** sometimes called Perma-Press or extra fluff, which allows your clothes to finish drying with the heat that has built up during the cycle instead of adding extra heat. Doing this will also help get rid of any heat-set wrinkles.

- **Dry your heavier cottons (such as towels) separately** from your lighter-weight fabrics—that way all the clothes in the machine will be dry at the same time. Also, don't add wet clothes to a load that's already partly dry. It will only end up taking longer than two shorter cycles would.

No Cost — More Savings, Less Carbon Dioxide

Annual amount of *money* **saved** as a result of always washing clothes in cold water and using the moisture-sensor setting on your dryer:	Annual amount of CO_2 **not** *emitted* as a result of always washing clothes in the cold water and using the moisture-sensor setting on your dryer:
$76	**1,533 lbs.**

Assumptions: Washer uses 1,542 kilowatt-hours per year (including hot water), dryer uses 1,090 kilowatt-hours per year, at 8.16 cents per kilowatt hour; washing in cold water reduces washing energy use by 50 percent; using the moisture-sensor setting reduces drying energy use by 15 percent.

Search for More Info

- www.pge.com/003_save_energy/003b_bus/003b1d2a_cal—
  This easy-to-use on-line calculator lets you see how much
  you'll save by washing in cold water versus hot water. Select
  "Laundry" from the pull-down menu.

- www.dulley.com/gtopics.shtml—Click on "Appliances
  (Other)," then scroll down to "Bulletin 621" to read all
  about why and how you should check your dryer's exhaust
  hose or vent.

- www.buildinggreen.com/products/washers.html—Here is a
  great introduction to H-axis washing machines. Also, see Tip
  14 in this book for more information.

You're Such a Dish

Selecting the air-dry setting (or opening the door and letting the dishes air dry) can cut your dishwasher's energy costs by 40 percent.

Overview. Whether or not you have a dishwasher, keep reading. No matter how you wash dishes, energy is needed to heat and pump the water you use. A dishwasher also needs electricity to run the machine. And even though washing dishes by hand may not guzzle as much energy as your clothes washer or fridge, it does guzzle as much as your dryer—about 1,000 kilowatt-hours a year! The good news is that there are simple ways to make your dishwasher (or dishwashing method) more energy-efficient—which means you can help conserve energy, water, *and* money.

What You Should Know

- A dishwasher uses the same amount of water and electricity whether it's half full or completely full.

- If you use two side-by-side sinks to wash your dishes by hand (one filled with hot, soapy water for washing and the other with cold water for rinsing), you'll use one-half as much water as you would using a dishwasher.

- Be careful, though. Letting the water run continuously while you wash dishes by hand for seven minutes uses approximately 17.5 gallons of water. Meanwhile, even the oldest, most inefficient dishwashers use only 15 gallons per load.

- Eighty percent of the energy consumed by your dishwasher is used to heat the water to 140°F, hotter than any other water in your home.

- ***Climate Results:*** If just once a year every household in America hand-washed their dishes using wash and rinse basins instead of letting the water run continuously, we would emit 324,000 fewer *tons* of carbon dioxide and save more than 1.2 *billion gallons* of water.

- ***Money Matters:*** Letting warm water run for seven minutes as you *pre*rinse dishes before putting them in the dishwasher will cost you around $60 a year.

Easy Ways You Can Help

- ***Don't prerinse dishes*** before putting them in the dishwasher! Today's detergents are designed to clean the dirtiest of plates. Scrape (don't rinse) off large pieces of food from your dishes, and experiment with your dishwasher to find out how much you really need to prerinse your dishes and still have them come out clean—you may be surprised!

Instead of scrubbing at tough pans while you let the water run, soak them overnight in warm water with a bit of soap. If you absolutely must prerinse, fill up a *cold*-water washbasin and, for energy's sake, don't let the water run continuously. Last, if you have a garbage disposal, always run it with *cold* water.

- **Always run your dishwasher with a full load.** Just make sure you don't overload it so much that the dishes don't get clean (again, you may need to experiment a bit to figure this out).

- **Press the "energy saver" or "light wash" switch.** This option uses less water, a shorter rinse cycle, and cold air to dry the dishes. If your machine doesn't have one of these switches, use the shortest cycle possible to get your dishes clean.

- **Select the air-dry option,** which is sometimes separate from the energy saver switch. Federal law requires that all new dishwashers have this option, which circulates air with the help of an internal fan to dry the dishes instead of pumping in new, hot air. It takes a little longer, but it can reduce your energy usage by 40 percent. If your older machine doesn't have this option, stop the machine before the drying cycle begins and open the door to let the dishes air-dry. Make a note of the time the first time you do this so you know exactly when to stop it in the future.

- **Try not to use the "rinse hold" option,** especially if you have only a few dirty dishes. This option rinses your dishes forcefully with large amounts of water, consuming three to seven extra gallons of hot water each time you use it. Try soaking those tough dishes instead.

- **Upgrade and save.** Investing in a new, energy-efficient dishwasher could save you up to 2,800 gallons of water and $44 a year (even after taking into account the cost of the new dishwasher). See Tip 14 for more information.

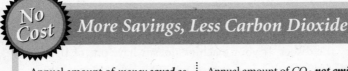

No Cost

More Savings, Less Carbon Dioxide

Annual amount of *money **saved*** as a result of using the air-dry setting on your dishwasher (or opening the door and letting the dishes air-dry):

$33

Annual amount of CO_2 ***not emitted*** as a result of using the air-dry setting on your dishwasher (or opening the door and letting the dishes air-dry):

663 lbs.

Assumption: Based on average U.S. dishwasher usage of 1,010 kilowatt-hours per year, at 8.16 cents per kilowatt-hour; air drying reduces energy usage by 40 percent.

Search For More Info

- www.eren.doe.gov/buildings/consumer_information/dish-wash/dishques.html—The Department of Energy answers your most frequently asked questions about dishwasher efficiency.

- www.sears.com—Click on "Appliances" and then on "Buying Guide." Select "Dishwashers" to learn all about what to look for in a new dishwasher.

- www.dulley.com/gtopics.shtml—Click on "Appliances (Kitchen)" and then scroll to bulletin 538 to read about the latest in quiet, efficient dishwashers.

- *Water Wars*, by Diane Raines Ward. Check out this book about the Earth's precious water supply and why we're killing it.

Winch the Water Heater

If half the households in America turned down their water heaters by 10°F, we could prevent 239 million tons of carbon dioxide from being emitted each year.

Overview. You have a thief living in your basement. Most Americans don't realize that, in terms of electricity, their water heater is the second greediest appliance in the house, next in line only to the fridge. Luckily, with a few simple changes you can tame this energy robber, making your water heater more efficient while still being able to enjoy a hot shower. Think it's impossible? Think you can't do it *and* save money? Think again.

What You Should Know

- Most people's water heaters are set to 140°F without them even knowing it. But 120°F is *plenty* hot for a shower and also hot enough to kill any bacteria that could build up in your heater—plus, at this lower setting you'll save *a lot* of energy!

- For every 10°F you turn down your water heater, you will save at least 5 percent of the energy it currently uses.

- Up to 30 percent of the energy used by your water heater goes toward keeping a huge tank full of water hot at all times. You can prevent this unnecessary waste of energy by installing a time-control switch or a heat trap (see the following section).

- *Climate Results:* Simply wrapping your water heater with an insulating jacket can save 5 to 10 percent of the energy typically used by the heater, or approximately 465 pounds of carbon dioxide a year.

- *Money Matters:* Wrapping your water heater *and* insulating any exposed hot-water pipes can reduce your annual hot-water bill by 15 percent. The cost of the wrap and insulation will be earned back in savings within one year.

Easy Ways You Can Help

- *Turn it down.* Turn the temperature of your water heater down to 120°F (or turn it to the "energy conservation" setting, if there is one). If you have an electric water heater with an upper and a lower thermostat, lower them both to 120°F. Also, check to see if your dishwasher has a booster heater. These internal water heaters bring the water temperature up to 140°F, allowing you to get rid of tough grease while keeping the water heater for the rest of your house set at 120°F. Look in your owner's manual to see if your dishwasher has one of these. If it doesn't, you may want to leave your home's water heater set at 140°F.

- **Bundle it up.** Wrap your water heater with a premade insulating jacket, available at your local hardware store for around $10. This will help prevent your tank from losing heat and reduce your annual water-heating costs by up to 10 percent. A wrap is especially important if your heater is either located in an unheated area (like a basement) or more than 10 years old. Be careful that the jacket does not block any air vents or cover the thermostat or flue. If you have a relatively new water heater, it may have come installed with an insulating jacket.

- **Insulate hot-water pipes** that exit your water heater for at least the first five to seven feet (keeping the insulation away from any gas flues or pilot lights). This insulation prevents heat from being lost as the water travels from your water heater to the rest of your house. Also, insulate any exposed hot-water pipes that are in crawl spaces or other unheated areas. You can buy easy-to-slip-on foam sleeves at your local hardware store for around two dollars each.

- **Get a gadget that preheats your water.** With the hot-water D'mand system, you can press a button near your shower, wait 30 seconds, and then turn on a blast of instantly steaming water. Think of how much water normally goes down the drain while you wait for the water to heat up! This super-efficient invention saves between $160 to $200 a year, can easily be added to an existing home, and will pay for itself within three years (see web site in the following section).

- **Have a heat trap installed.** If you can't afford a D'mand system right now, having a contractor install a $30 heat trap is another way to prevent wasting hot water, although the savings won't be as significant. A heat trap attaches to the pipe leaving your water heater and allows hot water to circulate through your tank when no one is using it (instead of letting it rise into the pipes, where it will eventually cool down and need to be reheated). A heat trap will reduce your annual water heating costs by 5 to 10 percent.

- **Install a timer.** By installing a timer on your water heater, you can set it to automatically turn off late at night and on again half an hour before you wake up. This simple installation will pay for itself within a year, and you can pick up a timer at your local hardware store.

- **Drain one quart of water** from the valve faucet at the bottom of your tank every three months. This procedure will make your heater more efficient by preventing sediment from building up.

- **Get a new one.** Because water heaters use so much electricity and last up to 15 years, having an energy-efficient one could mean substantial energy savings. If your water heater is more than seven years old, it may make more sense for you to upgrade now instead of waiting until it breaks. See Tip 16 for more information.

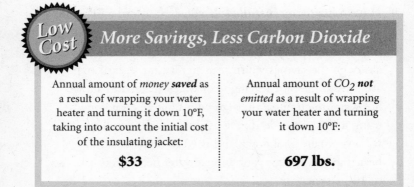

Low Cost

More Savings, Less Carbon Dioxide

Annual amount of *money **saved*** as a result of wrapping your water heater and turning it down 10°F, taking into account the initial cost of the insulating jacket:

$33

Annual amount of CO_2 ***not** emitted* as a result of wrapping your water heater and turning it down 10°F:

697 lbs.

Assumption: Based on the average U.S. household's consumption of 2,835 kilowatt-hours per year on water heating; wrapping the water heater reduces energy consumption by 10 percent; turning it down 10°F reduces energy consumption by 5 percent; an insulating wrap costs $10; measures done on a 6-year-old heater with a 12-year life span.

Search For More Info

- www.howstuffworks.com/water-heater.htm—Check out this web site to learn exactly how your water heater works.

- www.metlund.com—Go here to learn all about D'mand hot-water preheaters.

- www.homestore.com/HomeGarden/HomeImprovement/HowTos/HowTos—Here is a do-it-yourself guide on insulating your water heater. Click on "Plumbing" then "Insulating your water heater."

Shower Simply

Showering is responsible for 37 percent of a household's hot-water usage. The next biggest consumer, the washing machine, uses 26 percent.

Overview. Sure, taking a long, hot shower is luxuriating—but it also uses a lot of energy. Your shower is the biggest user of hot water in the entire house. Luckily for you, there's an inexpensive way to cut down on your shower's water usage (and it's not taking a shower once a week!). Allow us introduce you to a great invention called the low-flow showerhead. Because it adds air to the water, you'll have the same amount of water pressure as you did before, and you won't notice that less actual water is pouring down on you. A faucet aerator applies the same principle to your sink. We still want you to shower (your friends do too)—we just want you to learn how to shower with less energy.

What You Should Know

- Solely at the sink and in the shower, the average person uses 50 gallons of hot water a day.

- Older showerheads use 4 to 6 gallons of hot water per minute, but under new federal guidelines, any showerhead manufactured after 1994 must have a maximum flow rate of 2.5 gallons per minute. Meanwhile, an efficient low-flow showerhead uses only 1.0 to 2.5 gallons per minute. That means a low-flow showerhead could reduce your hot-water usage by up to 83 percent!

- If you're wondering how many gallons of water per minute your showerhead or faucet uses, turn the water on and time how long it takes to fill a gallon bucket.

- On average, a low-flow showerhead saves 15 gallons of water per shower compared to a standard showerhead.

- A typical sink faucet uses 3 to 4 gallons of water per minute, although federal guidelines require any faucets made after 1994 to have a maximum flow rate of 2.2 gallons per minute. A faucet aerator can reduce this to 0.5 gallons per minute! In other words, rinsing dishes for five minutes with a faucet aerator as opposed to an older faucet can save you up to 17.5 gallons of water!

- A laminar faucet, which sprays the water through many parallel streams instead of the whole faucet opening, greatly reduces water usage at your sink while rinsing just as forcefully.

- ***Climate Results:*** If only 3 percent of American households had one leaky faucet for one month, they would send 95 *million gallons* of water down the drain—that's enough to provide a year's supply of drinking water to 400,000 people.

- ***Money Matters:*** Putting faucet aerators on kitchen and bathroom sinks will save a typical family of three around $50 a year.

Easy Ways You Can Help

- *Fix leaky faucets.* A leak of just one drip every three seconds wastes 30 gallons of water a month! Finding your wrench and tightening the leak takes only five minutes.

- *Don't let the sink run* while you brush your teeth or shave. Letting hot water run without using it is one of the *worst* ways you can waste energy. A typical sink running for two minutes wastes at least six gallons of water, more if you have an older faucet. If shaving, plug and fill up the sink with water instead of letting it run continuously.

- *Take fewer baths.* Baths typically require 20 gallons of hot water, while a seven-minute shower with an efficient showerhead will use 14 gallons or less. Of course, a shower longer than 10 minutes almost always uses more hot water than a bath. Which brings us to our next tip . . .

- *Don't sing in the shower . . .* well, at least not for too long. A 5-minute shower is record speed, an 8-minute shower is normal, but anything above 10 minutes wastes a lot of hot water. An 8-minute shower compared with a 10-minute shower will save approximately 300 gallons of water a month. Sing while you get dressed instead—we're sure the whole house would love to hear you!

- *Install efficient showerheads.* Saving energy does not mean you have to shower under a trickle of water. Today's low-flow showerheads are not the same as yesterday's "flow restrictors." Modern low-flow showerheads reduce water flow while *adding in air.* The result is a strong spray that uses less water. Many utilities give these showerheads away for free or offer rebates on them. You can also pick one up at your hardware store for around $10 to $20 and easily install it yourself. Look for one with a shutoff valve, which allows you to turn off the water while you shampoo. When you turn it back on, the water will be at exactly the same temperature. Don't use a low-flow showerhead as an excuse to take a long shower, though. You won't save any energy or money that way!

- **Install faucet aerators.** A faucet aerator does the same thing as a low-flow showerhead: It mixes air into the water, resulting in a powerful stream that uses less water. They're two to five dollars each, and you can find them at your local hardware store. They're even simpler to install than low-flow showerheads—simply screw them on! For the bathroom sink, get an aerator that flows at 0.5 to 1.0 gallon per minute. For the kitchen sink, you'll need a bit more pressure, so get one that flows at 2 gallons per minute. If you wash dishes by hand, be sure to look for an aerator with a shutoff valve.

Search For More Info

- www.bae.ncsu.edu/programs/extension/publicat/wqwm/ he251. html—this web site shows you how to calculate exactly how much faucet aerators and low-flow showerheads can save *your* family.

Low Cost

More Savings, Less Carbon Dioxide

Annual amount of *money **saved*** as a result of installing 4 faucet aerators and 2 low-flow showerheads, taking into account initial cost of the new equipment:	Annual amount of CO_2 *not emitted* as a result of installing 4 faucet aerators and 2 low-flow shower-heads:	Annual amount of *water **saved*** as a result of installing 4 faucet aerators and 2 low-flow showerheads:
$255	**1,671 lbs.**	**10,950 gals.**

Assumptions: A three-person family, each person uses sinks five minutes a day and takes one eight-minute shower a day; old faucet is 4 gallons per minute; aerator makes it 2 gallons per minute; old showerhead is five gallons per minute, low-flow makes it 2.5 gallons per minute; cold water costs $1.50 per 1,000 gallons, electricity costs 8.16 cents per kilowatt-hour; 25 percent of water used at faucet and 60 percent of water at shower is heated; 0.2 kilowatt-hours used to heat 1 gallon of water from 50°F to 130°F; low-flow showerheads cost $15; faucet aerators cost $3.50.

- www.homestore.com/HomeGarden/HomeImprovement/ HowTos/HowTos—Here are detailed instructions on how to install a low-flow showerhead. Click on "Plumbing," then "Replacing a Shower Head."

- www.energyguide.com/gear/subdept.asp?dept%5Fid=23— Go here to buy low-flow showerheads and faucet aerators on-line.

Troubling Toilets

Forty percent of the pure, drinkable water your family uses each month goes to flushing your toilets, making them the single largest user of water in your home.

Overview. We don't think twice about them. In fact, it's only when we're unfortunate enough to have to hunt down the plunger that we actually appreciate our toilets. Just think of how amazing they are: With a simple push of a lever, our waste disappears from sight—*forever*. Toilets are indeed a great luxury—which is why we shouldn't be too greedy about them. In many older toilets, 5 to 7 gallons of drinkable water disappears down the drain when only 1½ gallons are actually needed to do the flushing. And remember, for every gallon of water you use, electricity is needed to pump that water through your pipes—and electricity used means carbon dioxide emitted at the power plant!

What You Should Know

- Over the past 30 years, the U.S. population has grown by 52 percent while water use has increased by 300 percent. Meanwhile, more than one billion people in this world don't have consistent access to drinking water.

- A toilet's gallon per flush (gpf) rating is usually painted somewhere on the toilet itself, such as in the space between the bowl and the tank. Check to see how many gallons per flush your toilet uses!

- In the 1960s toilets were manufactured to use 5½ to 7½ gallons per flush. In the 1970s some water-conscious citizens started demanding lower-flush toilets, which resulted in hastily designed 3½-gallon-per-flush toilets that didn't work so well. In the 1980s, the American National Standards Institute (ANSI) decided to make sure these 3.5-gallon-per-flush toilets worked properly. In 1994 ANSI convinced the government to require all new toilets to use only 1.6 gallons per flush. Be careful, though—some 1.6 gallons per flush toilets work better than others (see the following section).

- ***Climate Results:*** If 100,000 people each replaced just one 5½-gallon-per-flush toilet with a low-flow, 1.6-gallon-per-flush toilet, they would collectively save 702 *million gallons* of water and prevent 30,000 *tons* of carbon dioxide from being emitted each year.

- ***Money Matters:*** Ever since the U.S. government required all new toilets to use no more than 1.6 gallons of water per flush in 1994, Americans have saved *four billion dollars* a year on their water bills.

Easy Ways You Can Help

- ***Displace and save.*** If you have an older, 5½- to 7½-gallons per flush toilet but can't afford a new low-flow toilet, don't despair. Simply fill up a small plastic milk jug with water (put some stones in it to make it sink), and stick it in the tank of your toilet, making sure that it doesn't interfere with the flushing mechanism. The jug takes up space so not as much water can fit in the tank. As a result, one to two gallons *less* water is used per flush. You can also pick up a premade water displacer or "toilet dam" at your local hardware store.

- *Get an early-closure flapper.* Installing an inexpensive flapper is another way you can save water in older toilets. The flapper closes slightly early, giving enough time to flush out the waste but not the whole tank of water. The water pressure won't be affected, you can install an early-closure flapper yourself, and you'll save 30 to 50 percent of water per flush. One popular brand is the Frugal Flush Flapper.

- *Check for leaks.* Speaking of flaps, yours is probably leaking, which means water is continuously flowing down your toilet. Pick up some blue leak-detection tablets at your hardware store and drop them in your tank. If the water in your toilet bowl turns blue, you have a leak. Getting a new flapper (or better yet, an early-closure flapper) should fix it.

- *Go with the low flow.* If you have older toilets that use more water per flush than is necessary, consider replacing them with new, 1.6-gallons-per-flush low-flow toilets. Although all new toilets are required to use 1.6 gallons of water per flush, some work better than others, so you need to choose carefully. Ones that rely only on gravity often have so little water pressure that you need to flush twice—which defeats the whole purpose of having a low-flow toilet! We recommend getting a low-flow toilet that has added air pressure, providing a more forceful flush while using the same amount of water. Call your local water utility and ask if they will give you a rebate on a new low-flow toilet. Some utilities even pay for the replacement of an old toilet!

- *Select-a-flush.* If you're buying a new toilet, also keep this in mind: One of the latest inventions in toilets is the two-flush system. One button, for liquids, uses only 1.1 gallons per flush, while the other, for solids, uses the standard 1.6 gallons per flush. Since 90 percent of toilet flushes are for liquids only, we think this is a marvelous idea.

- *Shop for the boys.* If you're in the market for a new toilet and have males in your household, consider buying a hide-away flushless urinal. It's recessed into your wall and discreetly covered by a door when not in use. The best part is that it uses absolutely no water and doesn't smell! The patented EcoTrap traps the urine immediately and stops it from releasing any odor back into the room—all you need is a simple drain line (no water supply or flush valve needed). This wonderful invention is springing up in public rest rooms across the country—why not bring it into your home and save at least 15,000 gallons of water per year?

- *Shop for the environment.* The absolute *best* toilet you can buy in terms of saving water and energy is a composting toilet, which requires little to no water. This toilet flushes waste into an odor-sealed chamber, where it then composts the waste into natural fertilizer. Two to three times a year you empty out a small amount of dry, odorless, natural fertilizer, which you can apply directly to your garden or lawn. These toilets are completely clean, safe, and odor-free, and can save you more than 40,000 gallons of water per year!

No Cost

More Savings, Less Carbon Dioxide

Annual amount of *money **saved*** as a result of putting a water-filled milk jug into two 3.5-gpf toilets' tanks to reduce them to 1.6 gpf:	Annual amount of CO_2 ***not** emitted* as a result of putting a water-filled milk jug into two 3.5-gpf toilets' tanks to reduce them to 1.6 gpf:	Annual amount of *water **saved*** as a result of putting a water-filled jug into two 3.5-gpf toilets' tanks to reduce them to 1.6 gpf:
$35	**581 lbs.**	**6,840 gals.**

Assumptions: Both toilets are flushed five times per day, 365 days a year; water costs $0.005 per gallon; switching one toilet from 3.5 to 1.6 gallons per flush saves 177 kilowatt-hours a year.

Search For More Info

- www.terrylove.com/crtoilet.htm—This article ranks the best low-flow toilets on the market.

- www.waterless.com/Ecotrap.htm—Go here to learn how flushless urinals work.

- www.solareco.com/articles/article.cfm?id=100—Go here to learn all about composting toilets. You can also order a composting toilet directly from the web site.

Got Evian?

The electricity consumed by America's watercoolers in one year is equal to the amount of carbon dioxide emitted by 700,000 cars.

Overview. Ten years ago you might have laughed if someone tried to charge you $1.50 for water. Today, 12 ounces of spring water can cost you upwards of $3.00! Market analysts are baffled by our H_2O obsession—and kicking themselves for not having bought more Evian stock. Why are Americans drinking so much bottled water? Well, for starters, it tastes better. We also think it has fewer chemicals (although that's not always the case!), we want to lose weight and hydrate our bodies, water's better for us than soda, and most of all, we can afford to. What we don't realize, however, is that springwater takes a lot of energy to collect, bottle, and transport—much more than tap water. Luckily, there are ways to enjoy great-tasting water without the extra cost and energy.

What You Should Know

- The majority of the human population has to walk three or more hours to find drinkable water.

- According to a UCLA study, tap water in the United States is at least as safe as, if not more safe than, bottled water.

- You can drink 40,000 eight-ounce glasses of tap water for the price of three $1.50 bottles of springwater.

- Twenty-five to 40 percent of all bottled water in the United States, including some major brands, is made from tap water.

- An Energy Star watercooler with a low standby mode can save you $50 a year compared with a non–Energy Star cooler.

- Americans spend $300 million a year just to operate their water coolers.

- *Climate Results:* An instant hot-water dispenser is at least 20 percent more efficient than boiling water on a stove and can be up to 80 percent more efficient if you use it often. That means if 100,000 people used an instant hot-water dispenser instead of a stove to heat up just four cups of water twice a day, they would prevent *at least* 6,300 *tons* of carbon dioxide from being emitted each year.

- *Money Matters:* If you currently have a watercooler and go through at least one five-gallon jug every two weeks, you could save $175 or more a year by switching to a faucet filter.

Easy Ways You Can Help

- *Get a portable filtered bottle!* Not only are two-dollar plastic bottles of springwater a rip-off but they also fill up our landfills (since not everyone recycles the bottles), create air pollution (from the trucks that transport the bottles), and aren't always chemical- or bacteria-free. Instead, buy a 20-ounce portable sports bottle with a replaceable filter, such as Brita's Fill and Go bottle. Just fill the bottle with tap water, let the filter do its magic, and you're ready to go! Alternatively, you could refill a regular plastic water bottle at your sink's filtered faucet (see the following section).

- *Stay local.* If you must buy a bottle of water, buy the brand that is bottled nearest you. If the water's source is near you, it won't have to travel as many miles in a carbon-dioxide-emitting truck to reach your store. Check the back of the label to find out the water's source.

- *Ditch the watercooler . . .* and get a faucet filter! A faucet filter, such as Brita's new Ultra Faucet Filter, clips easily onto your faucet to give you filtered water right out of your tap. This is the most energy-efficient way for you to drink filtered water, since the tap water is being purified at a plant anyway. A faucet filter is even more efficient than a filtered water pitcher that you put in the fridge, since when you open the fridge door and cold air escapes, your fridge has to work harder to replace it. Look for a faucet filter with an indicator that lights up when it's time to change the filter.

- *Go for an instant hot-water dispenser!* Are you willing to trade in your watercooler for a faucet filter but don't want to give up the instant hot-water lever? You're in luck! An instant hot-water dispenser installed on your sink provides the same instant hot water, but because there isn't a big watercooler to run in addition, you'll end up saving electricity.

Here's how it works: A mini, superhot electric water heater is installed under your sink and connected to a special spout that is separate from your regular tap faucet. Simply push the spout's lever and you'll instantly get near-boiling, 190°F water for your tea, coffee, soups, et cetera. A typical hot-water dispenser can provide 60 cups of steaming hot water per hour. You can buy one of these dispensers at a hardware store and install it yourself or have a contractor do it.

- *If you absolutely must have a watercooler . . .* make sure you get one with an Energy Star label. A watercooler that has an Energy Star label uses between 60 and 90 percent less energy than a regular cooler. That's because a typical, inefficient watercooler uses tons of electricity when it's standing by and waiting for you to use it, while an Energy Star cooler must go into a low-wattage standby mode (see web site in following section). Also, buy water only from a company that collects, cleans, and then reuses its jugs.

Search For More Info

- http://yosemite1.epa.gov/estar/consumers.nsf/content/watercooler.htm—Read about how much money and electricity an Energy Star watercooler can save you.

Low Cost

More Savings, Less Carbon Dioxide

Annual amount of *money **saved*** as a result of using a filtered water bottle filled with tap water instead of buying 2 bottles of springwater a week, taking into account the initial cost of the filtered bottle and its replacement filters:

$127

Annual amount of CO_2 ***not*** emitted as a result of using a filtered water bottle filled with tap water instead of buying 2 bottles of springwater a week:

580 lbs.

Assumptions: *One half-liter bottle of water costs $1.50; national average cost of tap water = $0.0015 per gallon; filtered water bottle costs $7.99 plus six $3.49 replacement filters per year; tanker truck that gets 7 miles per gallon and holds 20,000 0.5-liter bottles' worth of springwater travels 30 miles round trip from source to bottling plant; delivery truck that gets 7 miles per gallon and holds 20,000 0.5-liter bottles travels 250 miles round trip; 1 gallon gasoline burned = 23 pounds carbon dioxide emitted; manufacturing emits carbon dioxide equal to five times weight of product; waste disposal emits carbon dioxide equal to weight of product; one full 0.5-liter plastic bottle weighs 18.5 ounces; one empty 0.5-liter plastic bottle weighs 0.7 ounces, Brita filter and three replacement filters weigh 5 pounds; assume consumer drives 8 miles round trip to pick up filter, replacement filters, and 15 more pounds of products in a 23-mile-per-gallon vehicle.*

- http://shop.toohome.com/index.asp?nPage=36—Here is an informative article on instant hot-water dispensers, how to install them, and where to buy them on-line.

- www.brita.com/202i.html—Check out Brita's faucet filter, portable sports bottle, and more.

Shake 'n' Bake—Efficiently

A person can use 50 percent less energy than someone else cooking exactly the same meal.

Overview. If you can't stand the heat, get out of the kitchen—or just use your kitchen more wisely! Baking with an oven can generate a lot of heat and use a significant amount of energy. During the summertime this extra heat can make your air conditioner work harder to keep the house cool. The good news is that simply changing your baking habits can reduce your energy usage and monthly bill.

What You Should Know

- The average American household bakes with an oven 185 times a year.

- Electric ovens have to heat up approximately 35 pounds of steel and four cubic feet of air before they can start cooking the food. In fact, the food absorbs only about 6 percent of the total heat produced by the oven!

- An oven with a convection setting has an internal fan that circulates the hot air while baking, allowing the food to cook much faster and therefore saving energy.

- Natural gas ovens (with electric ignitions, not pilot lights) are almost twice as efficient as electric ovens because they use their fuel directly to cook the food. Electric ovens use electricity, which must be converted from coal into electricity at a power plant, then transported to your home. There are notable downsides to having gas stoves or ovens, though— natural gas fumes are combustible and therefore must be ventilated to the outside, which also takes energy.

- *Climate Results:* If every oven owner in America peeked at her or his dinner cooking in the oven *one* fewer time a year, we would prevent nearly 7,000 *tons* of carbon dioxide from entering the atmosphere—every year.

- *Money Matters:* An electric oven with a convection fan costs 30 percent less to operate than a regular electric oven. It makes financial sense to buy one, despite the higher price tag, since a convection oven will more than make up for that higher price in energy savings over the years.

Easy Ways You Can Help

- *Don't peek!* Every time you open your oven door to check on food, or lift a lid off a pot on the stove, as much as 25 percent of the heat escapes. Use an oven light, timer, and meat thermometer to monitor your food instead. On the stove, a lid on the pan will reduce cooking time by 65 percent. Try to use pots with clear glass lids so you can watch the food while it cooks.

- *Keep it clean.* Ovens, microwaves, and stove burners that are clean reflect heat better than dirty ones, which means your food will cook faster. This will save *you* some cooking time and the *planet* some carbon dioxide—so keep 'em sparkling!

- *Preheat less.* Preheat your oven only when the recipe specifically calls for preheating (for instance, with breads or pastries). Adjust the oven racks *before* you preheat so heat doesn't escape while you rearrange them. Also, turn the oven off five minutes before your food is done cooking—the built-up heat will finish cooking the meal.

- *Use glass or ceramic pans* in the oven instead of metal ones. They hold heat better. This means you can turn down the temperature about 25°F and the food will cook just as quickly.

- *Bake large meals.* Today's gigantic ovens are not efficient for cooking small or medium-size meals. Use the oven only to cook large meals or several dishes at once. Furthermore, since reheating food always requires less cooking time than starting from scratch, cook a large meal at the beginning of the week and reheat the leftovers as the week goes by.

- *Keep the air flowing.* Food cooks more quickly when hot air can circulate freely in your oven. For this reason, leave at least one inch between pans and don't place one pan directly underneath another—stagger them instead. Last, if you use foil to catch food that drips, lay the foil on a rack *below* the one your cookware is on instead of on the same rack.

- *Use the convection setting.* Speaking of keeping the air flowing, if your oven has a convection fan setting, always use it. This circulates the air, allowing you to reduce cooking time by approximately 25 percent!

- *Consider buying a self-cleaning oven.* Self-cleaning ovens may cost more initially, but they are better insulated than regular ovens, which means it will take less time and energy to cook your food. Use the cleaning feature immediately after you bake something so less energy will be needed to reach the cleaning temperature (somewhere around 850°F!).

No Cost

More Savings, Less Carbon Dioxide

Annual amount of *money* **saved** as a result of baking 75 percent of oven-cooked meals in glass or ceramic cookware and every week, baking two things at once instead of at different times:

$10

Annual amount of CO_2 **not** emitted as a result of baking 75 percent of oven-cooked meals in glass or ceramic cookware and, every week, baking two things at once instead of at different times:

185 lbs.

Assumptions: Average U.S. household bakes 185 oven-cooked meals per year; average oven-cooked meal bakes for 40 minutes; average electric oven without convection uses 2,700 watts.

Search For More Info

- www.geappliances.com/advantium/home.htm—Check out the latest in oven technology—these energy-efficient ovens cook eight times as fast as conventional ovens, and they double as microwaves.

- www.geappliances.com/shop/prdct/ckng_elec—Go here to find new convection ovens, self-cleaning ovens, and even warming drawers.

Now You're Cooking

A pressure cooker cooks food 10 times faster and uses up to 75 percent less energy than a conventional oven.

Overview. Most people think that the only way to use less energy in the kitchen is either to buy a new, energy-efficient stove or to eat less. That's where they're wrong. Improving your cooking habits is *the* best way to conserve energy in the kitchen. The way a person cooks a meal can be the difference between an energy hog and an energy saint. A little letting the water boil away here, a little leaving the exhaust fan on too long there—these things may *seem* insignificant, but they can add up! Luckily, they're also very simple to change.

What You Should Know

- A microwave is approximately *75 percent* more energy-efficient than an electric oven. By increasing your use of the microwave for reheating, defrosting, browning, and cooking vegetables and popcorn, you can cut your total cooking costs by over 20 percent.

- Cutting up a steak into cubes will reduce its cooking time by 33 percent. A carrot that is cut up will cook twice as fast as a whole carrot.

- A toaster oven uses two-thirds less energy than a conventional electric oven.

- In terms of energy, microwaves are ideal for cooking and reheating food. But be careful! A microwave dinner uses more energy than a dinner made from scratch when you take into account the energy that was needed to process, precook, package, and transport that meal.

- *Climate Results:* Cooking with a six-inch pan on an eight-inch burner wastes over 40 percent of the burner's energy. If, just once, every household in America cooked with a pan that was the same size as its burner (as opposed to a pan that was smaller than its burner), we would prevent 30,000 *tons* of carbon dioxide from being emitted into the atmosphere.

- *Money Matters:* It costs two cents to cook a potato in a microwave but ten cents to bake one in an oven.

Easy Ways You Can Help

- *Smaller is better.* As a general rule, smaller appliances use less energy than larger appliances. Try always to use a pressure cooker, Crock-Pot, toaster oven, microwave, or stove (in that order) before resorting to the oven—especially if you're cooking smaller meals.

- **Watch water boil.** Once water (or any liquid) begins to boil, it won't get any hotter. When your water starts to boil, turn down your stove's burner a bit—you'll maintain the boil *and* save some energy. Also, don't boil any more water than you need—it will just take longer, and therefore use more energy.

- **Size it right.** Always match the size of the pan you're cooking with to the size of the burner—that way less heat will be lost to the surrounding air.

- **Play with knives.** Cut up your food into smaller pieces before cooking—it will cook up to twice as fast!

- **Thaw it right.** If you can plan your meals enough hours in advance, thaw your frozen food in the refrigerator. This will give your fridge a break since the frozen food will help to cool the other food in the fridge. If you need dinner *now*, thaw it in the microwave instead of under hot water—the microwave will be much more energy-efficient.

- **Cover liquids and wrap foods in your fridge,** since uncovered liquids and foods release moisture, making it harder for your fridge to cool down. Also, wrap foods with aluminum foil or plastic wrap instead of wax paper. Paper acts as an insulator, so it keeps your food warm longer.

- **Keep it cool.** When you're having company over, fill a cooler or bucket with ice so your guests don't have to keep opening and closing the freezer (and letting in warm air). In the summertime, fill a pitcher with tap water and keep it in the fridge so you don't have to let the tap water run while you wait for it to get cold.

- **Let out some steam.** Turn on the exhaust fan above your stove whenever your food is creating unwanted heat and steam. In the wintertime, though, this heat and steam can help heat your house (but if the steam smells like garlic, we understand why you'd want to use the exhaust fan!). In the summertime, it's always a good idea to vent that hot air outside before it steams up your house and makes your air conditioner work harder.

  No matter what the season, don't leave the exhaust fan on for too long. Once the fan pulls the hot air away from your stove, it will begin to suck up the air from your house. In the wintertime, an exhaust fan left on high can suck out an entire roomful of heated air in just four minutes, making your furnace work double time to replace it. In the summer, your precious air-conditioned air will be sucked directly outside if you leave the fan on too long. If your food needs to boil for an extended period, leave the fan on as low as possible.

- **Know your stove.** If you have an electric stove, use only flat-bottomed pans—rounded pans will lose most of their heat. If you have a gas stove, make sure the flame is blue, which means the gas is burning efficiently (a yellowish flame indicates that your pilot light needs adjusting). And no matter what kind of stove you have, know that copper pans heat up faster than any other type.

No Cost — *More Savings, Less Carbon Dioxide*

Annual amount of *money **saved*** as a result of cooking ⅓ of oven-baked meals in a microwave, ⅓ on the stove, and ⅓ in the oven: **$13**	Annual amount of CO_2 ***not*** *emitted* as a result of cooking ⅓ of oven-baked meals in a microwave, ⅓ on the stove, and ⅓ in the oven: **254 lbs.**

Assumptions: Based on U.S. national average of 185 oven-cooked meals per household per year; average electric oven-cooked meal takes 40 minutes and uses 1.8 kilowatt-hours; average electric stove-cooked meal takes 35 minutes and uses 0.7765 kilowatt-hours; average microwave-cooked meal takes 22 minutes and uses 0.312 kilowatt-hours.

Search For More Info

- www.sears.com—To order a pressure cooker on-line, go to this page and search for "pressure cookers" in the box that says, "Search by Keyword or Item #." Also, here are some great pressure cooker recipes: www.gopresto.com/ppcinsti tute/recipes/pcrecipes.html.

- www.uniongas.com/NaturalGasInfo/NaturalGasSafety/main tenancetips.asp—Read this brief article about ventilation safety in the kitchen, especially if you have a natural gas stove or oven.

- www.modbee.com/life/healthyliving/story/1203042p-12705 23c.html—Check out these food safety do's and don'ts.

Phantom Loads

TV's and VCR's that are turned "off" cost Americans nearly a billion dollars a year in electricity.

Overview. *Off* does not necessarily mean "off." In other words, even when you turn off your TV, a small amount of electricity is being used for the remote-control feature. After one day, this won't add up to much. But after 24 hours a day, 365 days a year, it will! Furthermore, these phantom loads of electricity are quickly increasing as Americans get more and more gadget happy. Today's cell phone chargers, DVD players, satellite dishes, and digital TV's are quietly sucking up small amounts of electricity day in and day out. In fact, at least 5 percent of your annual electricity bill is consumed by these appliances while they're switched off. Surprised? Don't worry—most Americans don't know about this "leaking" electricity effect. But now that you do, you can do something about it.

What You Should Know

- One-fourth of the energy consumed each year by your TV is used when it's turned off.

- Any appliance or piece of electronic or office equipment that has a remote control, battery charger, internal memory, AC adapter plug, instant-on feature, permanent display (such as a clock), or sensor (such as a security alarm) will use electricity even when it's switched off. This includes microwaves, TV's, VCR's, cable boxes, garage door openers, cordless phones, battery chargers, stereo equipment, and much, much more!

- Some compact stereos (where the radio, tape deck, and CD player are all in the same housing) use 27 watts of electricity when they're on and 25 watts when they're off. Why even bother turning them off? The only way to stop wasting that energy is to unplug them.

- Many appliances such as ovens, VCR's, and microwaves have extra clocks on them. The clock itself needs only half a watt to run. However, the appliance is plugged into a 120-volt plug. With the exception of new models that have the latest in stand-by technology, trying to run a half-watt clock on a 120 volt plug is very inefficient. In fact, the appliance ends up using four to eight watts to run the half-watt clock. Just one day's worth of that wasted energy could power a compact fluorescent lightbulb for 10 hours!

- Wall cube plugs, like those on answering machines or cell phone chargers, are phantom loads in disguise. These plugs convert high-voltage alternating current (AC) from your outlet into low-voltage direct current (DC) for your appliance.

Even when the appliance is switched off, or finished charging, that little converter consumes 20 to 50 percent of the appliance's "on" wattage.

- *Climate Results:* If 100,000 people plugged in their VCR's only when they watched a movie, they would prevent more than 5,000 *tons* of carbon dioxide from being emitted every year.

- *Money Matters:* The average American household spends $40 a year on phantom loads, that is, electricity used when appliances are supposedly off.

Easy Ways You Can Help

- *Unplug!* This is the *only* way you can be absolutely sure that your appliance isn't using any energy. Certain appliances—coffeemakers, for instance—are not that hard to unplug after you're done using them. For most video and audio equipment, however, unplugging can be inconvenient. So, when you go on vacation for more than three days, take the extra few minutes to unplug your equipment. When you're not on vacation . . .

- *Use power strips.* Power strips do use a minuscule amount of electricity when they're plugged in. However, if you plug your TV, VCR, DVD player, and stereo system all into one power strip, switching the power strip off and dealing with its 1 to 3 watts of leaking electricity is much more energy-efficient than paying for the 75 watts of leaking electricity you would have if all those appliances were plugged in separately. The bottom line? Invest in a six-dollar power strip, plug your video and audio equipment into it, and switch it off every night. Don't forget that power strips have the added benefit of protecting your equipment from power surges.

- ***Buy equipment with low phantom loads.*** When you're looking for a new TV, VCR, or stereo system, remember that things like permanent clock displays or remote control devices are automatic phantom loads. Be sure to ask the store clerk how many watts the appliance uses when it's off and whether it has an automatic sleep/standby/low-power mode. The easiest way to know if a product has a low standby mode is to see if it carries the Energy Star logo (see Tip 15 for more info). Thanks to its low standby ratings, Energy Star video and audio equipment uses up to 75 percent less energy when turned off than does regular equipment.

- ***Batteries aren't better.*** Just because a lot of your appliances use electricity even when they're turned off you shouldn't be fooled into thinking that batteries are better! Disposable batteries (not rechargeable ones) are harmful to the environment because they contain toxic metals that are impossible to get rid of. Most disposable batteries get thrown into the garbage and sent to landfills, where they are crushed and thereby release their toxic metals into the air (see Tip 47). Even if they happen to be disposed of correctly, at a hazardous waste center, their toxic metals must be separated out and stored *forever*. Plus, manufacturing one disposable battery requires

50 times as much energy as that battery will provide when it's used. Last, the electricity you get from a battery is over *1,000* times more expensive than electricity from an outlet. Always try to buy appliances that plug into your wall and have a low standby rating. If you absolutely must buy an appliance that uses batteries, make sure it uses rechargeable ones.

- **The One-Watt Initiative.** The International Energy Agency is pushing for a Global One-Watt Plan, whereby all domestic appliances would be required to use one watt of electricity or less when they aren't on. The technology for one-watt standby power is available *right now*, but most manufacturers don't build their appliances with it because they don't feel the need. Only when the public starts to demand more energy-efficient products will the manufacturers listen. You can do your part by purchasing the absolute best Energy Star products there are—the ones that use one watt in standby mode.

More Savings, Less Carbon Dioxide

Annual amount of *money* **saved** as a result of plugging in your VCR and stereo only when you want to use them:	Annual amount of CO_2 **not** *emitted* as a result of plugging in your VCR and stereo only when you want to use them:
$12	**233 lbs.**

Assumptions: VCR is in 5.9-watt standby mode 72 percent of the time, 13-watt idle mode (on but not being used) 24 percent of the time; component stereo system is in 3-watt standby mode 65 percent of the time, 43-watt idle mode 16 percent of the time.

Search For More Info

- http://standby.lbl.gov/index.html—Check out the U.S. government's home page for standby power (also known as phantom loads or leaking electricity).

- www.earthinstitute.columbia.edu/library/earthmatters/ spring2000/pages/page25.html—This is a great introductory article on phantom loads of electricity.

- http://eetd.lbl.gov/EA/Reports/46212—This in-depth report will tell you exactly how much energy your electronic equipment uses each year.

You've Got (E-)Mail

It takes 99 percent more energy to manufacture a sheet of paper than it does to print a page out on a printer.

Overview. Today nearly 40 percent of American households own a personal computer. With the advent of e-mail, the Internet, chat rooms, and instant messaging, computer use is the fastest growing electrical load in the United States. Although a single computer doesn't consume that much energy, a printer, copier, fax machine, scanner, and CD burner certainly do. To make matters worse, Americans also have a nasty habit of leaving this equipment on for hours on end. While you're out enjoying your lunch break, your computer is sucking up electricity and contributing to global warming. As our culture grows more and more dependent on computers and the Internet, it is all the more important that we learn how to use these wonderful technologies efficiently.

What You Should Know

- Americans' computers and their monitors consumed 25 *billion* kilowatt-hours of electricity in 1999, making them second only to TV's in the use of energy by home electronic items.

- Today's computers are designed to handle 20,000 on-off cycles before their hard drives begin to wear down. That's equivalent to turning your computer on and off seven times a day for eight years.

- *Climate Results:* If all the people who owned printers in their home offices used just one ream (500 sheets) less of paper a year, whether through conservation or through more double-sided printing, they would prevent 130,000 *tons* of carbon dioxide from being emitted every year.

- *Money Matters:* A typical home computer and printer consume about $50 a year in electricity. Learning to use these machines efficiently can cut that cost in half.

Easy Ways You Can Help

- *Turn off your power strips!* Almost all electronic equipment continues to use a small amount of electricity even when it's turned off (see Tip 11). With a computer, printer, fax machine, and copier plugged in, this leaking electricity can turn into a genuine torrent of energy. For example, even if you leave these machines turned off for 18 hours a day, in a year they'll have consumed 40,000 watts during the time they were off. The only way to prevent this loss is to unplug these machines or turn off the power strip to which they are connected.

Computers

- ***Ditch that screen saver.*** Screen savers *do not* save energy—in fact, they use just as much energy as working on a spreadsheet does. They were invented to keep the image of your document from being "burned" into your monitor during periods of inactivity. Not only is burning no longer a concern for today's monitors, but an even better solution has come along: sleep mode.

- ***Put it to sleep.*** Check to see if your computer has a sleep mode, in which the screen goes dark after a certain period of inactivity. Sleep mode is a low-energy state that reduces electricity consumption by up to 80 percent. When you move the mouse, your computer will instantly "awaken" to where you left off, and you won't have to worry about losing unsaved information. To enable the sleep mode on your computer, go into the control panel and look for the power management folder. There, you should be able to set your computer and/or monitor to go on sleep mode (also known as standby) after a certain amount of time. You may need to restart your computer for the new settings to take effect.

- ***Better yet, turn it off.*** A computer in sleep mode still uses electricity. If you won't be using your computer for more than an hour, shut it all the way down. The notion that turning your computer on and off too many times will damage the hard drive is outdated. Today's hard drives are designed to handle 20,000 on-off cycles, as well as to withstand all the electrical and thermal surges that occur during start-up. In fact, frequently turning your computer off can actually *lengthen* its life span since you're using the hard drive less.

- ***Don't forget the monitor.*** A typical monitor screen uses much more electricity than the hard drive of a computer. You should also know that larger or higher-resolution monitors use more electricity than smaller, low-resolution ones. Laptop liquid crystal display (LCD) screens are by far the most efficient monitors; in fact, they're 80 to 90 percent more efficient than your typical cathode ray tube (CRT) monitor on a desktop computer. That's why some of the newer, flat-screen desktop monitors use LCD technology. The bottom line? First, buy the most efficient monitor you can afford, then turn off your monitor if you won't be using it for 15 minutes or more.

Printers and Copiers

- ***Turn that puppy off.*** Unless you have an Energy Star printer with a guaranteed low standby mode (see Tip 15), your printer uses the most energy while it's sitting idle. If you have a smaller printer, turn it *off* if you won't be using it for more than 15 minutes. If you have a larger printer that takes a long time to warm up, turn it off if you won't be using it for more than two hours.

- ***Copy all at once.*** A photocopier is typically the highest energy user in a home office, using roughly the same amount of energy when copying as your dishwasher. Copiers also use a lot of energy in standby mode—more than laptops do when they're in use. The best thing you can do to save energy is copy on both sides of the paper. Also, try to do all your copying or printing at once so you can leave the machine off for the rest of the day.

- *Use less paper.* Saving paper is the *best* thing you can do to conserve energy in your home office, since it takes so much energy to manufacture paper and transport it to stores, and then to your house. Here are some easy ways to conserve paper: Keep the spacing as small and the margins as narrow as possible on your documents. Use recycled paper. Print out only final copies (not drafts), print and copy on both sides of the paper when possible, and print on the back sides of scrap paper when you can. Send e-mail and electronic faxes instead of paper letters and faxes whenever possible; they will be quicker *and* save paper. Also, many companies now give you the choice to be billed and/or to pay your bills via e-mail and direct deposit. Electronic banking can be scary at first, but once you get the hang of it, you'll love its simplicity—and the environment will love the paper savings!

More Savings, Less Carbon Dioxide

Annual amount of *money* **saved** as a result of enabling the sleep feature on your computer, monitor, and printer to go on after 5 minutes of inactivity: **$22**	Annual amount of CO_2 **not** emitted as a result of enabling the sleep feature on your computer, monitor, and printer to go on after 5 minutes of inactivity: **435 lbs.**

Assumptions: Based on U.S. household average of 317 kilowatt-hours per year combined computer and monitor consumption and 250 kilowatt-hours per year printer consumption; computer sleeps in 15 watts; printer sleeps in 10 watts; both computer and printer used an average of 4 hours per day, 365 days a year, and sleep half of those hours.

Search For More Info

- www.eweb.org/energy/energysmart/edb/9906/computer.
  html—Check out this brief article on how turning off your
  computer and monitor will save you money without harming
  your hard drive.

- www.ns.ec.gc.ca/udo/office/office.html—This in-depth pam-
  phlet tells you how to make your office "green."

- www.tufts.edu/tie/tci/pcpowermanagement/FAQ.html—
  This site answers your most frequently asked questions about
  how to set your computer to go to sleep after a specified
  period of time.

Things Not Everyone Has

If every American who owned a pool or hot tub turned down its temperature by 4°F, they would prevent more than two billion tons of carbon dioxide from entering the atmosphere each year.

Overview. Ahhh, the romantic crackling of a fire in a fireplace. The bubbling of a hot tub on the back patio. The inviting shimmer of a pool on a hot summer day. The coziness of a warm water bed calling your name. There's no doubt about it—indulgences like these are wonderful. We mustn't be *too* greedy, though. These luxuries use a lot of energy, hurting not only our environment but also our pocketbooks. Luckily, you can improve the efficiency of these amenities with a few simple changes. Then you can sit back, relax, and know you've done something to lessen the effects of climate change.

What You Should Know

- A mere seven-mile-per-hour wind at the surface of a pool can increase its heat loss by *300* percent.

- During the wintertime, leaving the damper of your fireplace open when a fire is not going is like leaving a four-foot window wide open.

- The overwhelming majority of heat loss in a pool is caused by the natural evaporation of water at the surface. This evaporation requires *a lot* of energy. Here's an example: It takes nine kilowatt-hours of electricity for your pool heater to heat one pound of water from 50° to 80°F. Meanwhile, your pool loses *35 times* that amount of heat when the same pound of water evaporates into the air. Sound complicated? Don't worry, there's an easy solution: a pool cover.

- ***Climate Results:*** If everyone in America who owned a pool filter pump ran it for only four hours a day, they would prevent seven *million tons* of carbon dioxide from being emitted each year.

- ***Money Matters:*** If you never make your water bed and then start making it every day, you could save more than $30 a year in electricity.

Easy Ways You Can Help

Swimming Pools

- ***Cover up.*** A whopping 70 percent of the heat lost by outdoor pools is the result of water evaporation. Using a solar pool cover when you're not in the pool reduces this water loss by up to 50 percent. It also prevents overnight heat loss and helps keep your pool cleaner (which, incidentally, reduces the amount of chemicals you need to clean the water). With automatic reel systems that make covers easy to put on and take off, using a solar pool cover has never been easier.

- *Reduce your filter pump's operating time* to the shortest period that allows you to keep your water clean. Most pool pumps run *much* longer than necessary. As a general rule, one complete water turnover every 24 hours is plenty. This means running your filter pump only four to five hours a day, preferably during off-peak times (six in the morning till noon or at night) to conserve even more energy.

- *It's not a bathtub.* Lowering your pool's temperature by just 1°F can reduce your energy costs by 10 percent. Furthermore, the American Red Cross reports that the most healthful swimming temperature is 78°F. Turn your pool's temperature down as low as you can while still feeling comfortable. Last, if you won't be using your pool for five or more days, turn off the heater completely, including the pilot light.

- *Consider going solar* with your pool's heating system. It's well worth it—see Tip 17.

Hot Tubs

- *Give your hot tub a blanket.* A floating thermal blanket that goes underneath your solid hot tub cover will help retain heat, reducing your hot tub's energy usage by up to 33 percent. It will also preserve the life of your solid hot tub cover by reducing the moisture that builds up on the underside of the cover.

Water Beds

- *Make your bed.* A covered water bed retains 30 percent more heat than an uncovered one. In other words, a covered bed's water heater doesn't have to work as hard to keep the bed warm, and therefore uses less electricity. For even more heat retention, add extra blankets during the day.

- *Get a timer.* Plug your water bed into a timer so you don't have to leave it on all day. Set it according to your schedule, so the heater shuts off after you wake up and turns back on a half hour before you go to bed.

- *Thinking of buying a water bed?* Keep these things in mind: A soft-side water bed uses less energy than a hard-side frame. Also, the deeper the bed, the more energy will be needed to heat it. Last, a solid-state water bed heater uses 50 percent less energy than other heaters.

Fireplaces

- *Just for decoration?* If you never use your fireplace, be sure to seal the damper permanently with heat-resistant caulking. That way warm air from your home won't escape up and out of your chimney.

- **Always keep your damper closed** when a fire isn't burning; otherwise, the warm air in your house will fly right up the open chimney.

- **Turn your furnace down to 55°F** when using your fireplace. Unless you have a new clean-burning, advanced combustion fireplace, making a fire on a cold winter day almost always costs you more than using your heating system. That's because a large amount of heated air from your house goes up the chimney, which means your furnace has to work harder to replace it.

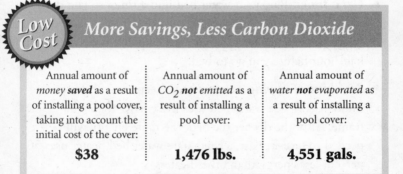

Low Cost

More Savings, Less Carbon Dioxide

Annual amount of *money **saved*** as a result of installing a pool cover, taking into account the initial cost of the cover:	Annual amount of CO_2 ***not*** *emitted* as a result of installing a pool cover:	Annual amount of *water **not** evaporated* as a result of installing a pool cover:
$38	**1,476 lbs.**	**4,551 gals.**

Assumptions: Based on average U.S. pool that's 16 by 32 by 5 feet and uses 3,000 kilowatt-hours per year; cover reduces energy use by 30 percent and water loss by 40 percent; cover costs $70 and lasts at least two years.

Search For More Info

- www.eren.doe.gov/rspec—Go to this site if you have a pool. Go now. This Department of Energy web site will tell you *everything* you need to know about making your pool more energy-efficient. Be sure to check out the "Energy Smart Pools Software" to calculate your potential energy savings.

- http://energy-publications.nrcan.gc.ca/index_e.cfm—Check out this article explaining why your fireplace is so inefficient. Scroll down and click on "Heating and Cooling" then scroll to the article entitled "All About Wood Fireplaces."

- www.pools.com/catalog—Go here and click on "Solar Covers" to buy them on-line. You can also buy a floating blanket for your hot tub by clicking on "Spas," then "Accessories."

Out with the Old, Save with the New

A typical refrigerator costs about $900. However, operating that refrigerator for 18 years (its expected lifetime) will cost you more than $2,000.

Overview. Upgrading can be a scary thing, especially when the most energy-efficient appliance is also the most expensive one. Don't panic. Energy-efficient appliances cost a lot to buy, but they will quickly pay for themselves and start saving you money, thanks to much lower operating costs. So, when you're out shopping for a new dishwasher, keep in mind that there are two price tags to consider: the initial cost of the dishwasher and the cost of operating it for the next 11 years. You may find that you'll save more money by upgrading *now* even if your current appliance isn't on its last legs. Most important, an energy-efficient appliance is guaranteed to save you electricity, which means less carbon dioxide will be emitted at the power plant.

What You Should Know

- The U.S. government has created two labels to help you identify energy-efficient products: the EnergyGuide label and the Energy Star logo.

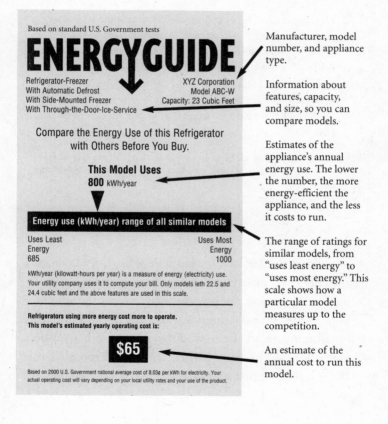

Based on standard U.S. Government tests

ENERGYGUIDE

Refrigerator-Freezer
With Automatic Defrost
With Side-Mounted Freezer
With Through-the-Door-Ice-Service

XYZ Corporation
Model ABC-W
Capacity: 23 Cubic Feet

Compare the Energy Use of this Refrigerator with Others Before You Buy.

This Model Uses
800 kWh/year

Energy use (kWh/year) range of all similar models

Uses Least
Energy
685

Uses Most
Energy
1000

kWh/year (kilowatt-hours per year) is a measure of energy (electricity) use. Your utility company uses it to compute your bill. Only models ieth 22.5 and 24.4 cubic feet and the above features are used in this scale.

Refrigerators using more energy cost more to operate.
This model's estimated yearly operating cost is:

$65

Based on 2000 U.S. Government national average cost of 8.03¢ per kWh for electricity. Your actual operating cost will vary depending on your local utility rates and your use of the product.

Manufacturer, model number, and appliance type.

Information about features, capacity, and size, so you can compare models.

Estimates of the appliance's annual energy use. The lower the number, the more energy-efficient the appliance, and the less it costs to run.

The range of ratings for similar models, from "uses least energy" to "uses most energy." This scale shows how a particular model measures up to the competition.

An estimate of the annual cost to run this model.

How to Find the Life-Cycle Cost of an Appliance

1. Look at your electric bill and see how much you pay per kilo-watt-hour (kWh) of electricity. Multiply that by the kilowatt-hours per year under "This Model Uses" on the EnergyGuide label to find the annual operating cost of the appliance.

2. Multiply the annual operating cost by the estimated number of years the appliance will last (see "What You Need to Know") and you will have the total operating cost over the life of the appliance.

3. Add to that the initial cost of the appliance, plus any installation or maintenance fees (ask the store clerk), and you will have the total life-cycle cost of the appliance.

4. Do Steps 1 through 3 for other brands and models to compare costs.

Note: EnergyGuide labels are based on an appliance's regular settings and don't reflect the use of any optional energy-saving or energy-guzzling features or cycles.

- The EnergyGuide label makes it easy for you to compare the energy efficiency of different brands and models of appliances, so you can buy the most efficient product for your needs. Manufacturers are *required* to put these labels on all refrigerators, freezers, dishwashers, clothes washers, water heaters, furnaces, boilers, central and room air conditioners, heat pumps, and pool heaters. So if you don't see it, ask for it!

- The Energy Star logo is a stamp of exceptional efficiency. It is awarded only to products that go above and beyond the minimum efficiency standards set by the government. If you see this logo, it means energy savings for the environment and money savings for you.

Money Isn't All You're Saving

- *Climate Results:* If just 20 percent of all Americans who owned washing machines upgraded to high-efficiency machines, they would prevent 3½ *million tons* of carbon dioxide from entering the atmosphere every year.

- *Money Matters:* There are many incentives—such as tax credits, cash rebates, special utility rates, and low-interest loans—out there for you to purchase energy-efficient products. Ask your salesperson, local utility, or local and state governments, and check out this web site (www.dsireusa.org) to find out which incentives are available in your area.

What You Need To Know

- *Refrigerators.* Your refrigerator is the biggest energy-using appliance in your home, and a new one will last you about 18 years, so it's *extremely* important that you take energy efficiency into account when buying a new fridge. Currently, the best Energy Star fridge uses less than half the energy used by a typical refrigerator, meaning it can cut your operating costs by more than $70 a year!

→ *Why They're Better.* In the last decade, refrigerator technology has improved faster than that of any other appliance. Today's energy-efficient fridges have better insulation, tighter door seals, more efficient compressors, and more accurate temperature controls. Refrigerated drawers that are separate from the fridge are also becoming popular. You can use these for specific foods such as fruits or vegetables. The drawers save energy because they let you open a single drawer instead of the entire refrigerator when you want something. Last, no refrigerators built after 1995 contain ozone-depleting chlorofluorocarbons in their coolant fluid.

→ *So Much to Choose From.* Feeling overwhelmed by all the choices in refrigerator features? Here are some things to keep in mind: Side-by-side refrigerator-freezers are 10 percent *less* efficient than ones with the freezer mounted above the fridge. Smaller fridges use less energy but may require more trips to the store, so pick the size that is just big enough for your household (this concept also applies to washing machines and dishwashers). Last, definitely choose a model that has automatic moisture control, which means it prevents condensation buildup without the help of a miniheater.

- *Washing Machines.* Your washer will last about 14 years, so you want to choose a new one wisely. A typical washer uses 40 to 60 gallons of water per load, while Energy Star washers use 50 percent less water and 70 percent less energy. That means an Energy Star washer could save you up to $100 and 7,500 gallons of water a year!

→ **Why They're Better.** Today's energy-efficient washers have a higher-speed spin cycle, removing more water from the clothes, which means your dryer won't have to run as long. They also have computer chips that sense the load size, fabric type, and dirtiness of the water and then adjust the water level, temperature, and type of cycle accordingly. Last, they use higher-pressure sprays during the rinse cycle, which get suds out more quickly and effectively. Studies have shown that these washers clean and rinse clothes better than standard washers while treating the clothes more gently.

→ **The Amazing H-Axis.** The most efficient washer you can buy today is called a horizontal-axis, or H-axis machine. These washers spin horizontally, just like your dryer, and tumble the clothes through a pool of shallow water. This design requires *much* less water than a standard washer, which soaks your clothes in a 40 to 60 gallon vat. H-axis washers are usually front-loading, just like your dryer, although there are some top-loading, H-axis washers on the market now.

- **Dishwashers.** Your dishwasher will probably last for 11 years, so upgrading to an Energy Star model that is 25 percent more efficient and saves up to 2,300 gallons of water per year may be worth your while.

  → **Why They're Better.** Since 80 percent of the energy consumed by your dishwasher is used to heat the water, recent technology has focused on reducing the amount of water needed for each load. Energy Star dishwashers have better spray arms that clean dishes faster. They also have sensors that can tell how dirty your dishes are and adjust the length, type, and temperature of the cycle accordingly. Almost all new dishwashers have internal booster heaters that raise the temperature of the water to 140°F, allowing you to turn your home's water heater down to 120°F (see Tip 5). Last, make sure you choose a model that has an air-dry setting and energy-saving options such as "light wash" or "energy-saver wash."

Easy Ways You Can Help

- *Upgrade!* First, figure out how much your current appliances are costing you and research the most efficient ones on the market. See if upgrading will save you money (as well as energy). Once you've decided to upgrade, follow these five simple steps.

  1. Determine how much you can afford to spend on a new appliance, taking into account any rebates or incentives being offered by your utility or local or state government.
  2. Consider only appliances that have the Energy Star logo.
  3. Decide what you want in features and size.
  4. Use the EnergyGuide label to figure out the appliance's "second" price tag (the operating cost over its expected life span). Then figure out the second price tags of different brands and models so you can compare them.
  5. Choose the most energy-efficient model that meets your needs.

Investment

More Savings, Less Carbon Dioxide

Annual amount of *money **saved** and carbon dioxide **not** emitted*, as well as the total amount of money saved over the life the appliance, as a result of upgrading to an Energy Star model, taking into account the initial cost of the new, Energy Star appliance:

Energy Star Refrigerator:	*Energy Star Washing Machine:*	*Energy Star Dishwasher:*
$32/Yr.	**$30**/Yr.	**$15**/Yr.
1,486 lbs. CO_2/yr.	**1,055** lbs. CO_2/yr.	**756** lbs. CO_2/yr.
$576/over 18 yrs.	**$420**/over 14 yrs.	**$165**/over 11 yrs.

Assumptions: *Upgrading from a typical 1,323-kilowatt-hour-per-year refrigerator, 1,397-kilowatt-hour-per-year washing machine, and 1,010-kilowatt-hour-per-year dishwasher to a 417 kilowatt-hour-per-year Energy Star refrigerator costing $750, a 466-kilowatt-hour-per-year Energy Star washer costing $650, and a 549-kilowatt-hour-per-year Energy Star dishwasher that costs $250; all dishwasher and washing machine kilowatt-hours include hot water.*

Search For More Info

- www.energystar.gov/products/appliances.shtml—Start here to learn all about Energy Star appliances, how much they can save you, and where you can buy them.

- http://aceee.org/consumerguide/mostenef.htm—You might want to cross-check the appliances listed on the Energy Star web site with the ones listed here under "The Most Energy-Efficient Appliances of 2001."

- www.energystar.gov/stores/storelocator.asp—Type in your zip code to find a store that sells Energy Star appliances near you, or sells them on-line.

All I Want for Christmas Is a New Laptop

Laptops are 90 percent more energy-efficient than desktop computers.

Overview. After using less paper, the next best thing you can do to save energy in your home office is to use energy-efficient equipment. If your equipment is nearing five years of age, consider making the investment to buy a more efficient model. It might save you more money in the long run. Furthermore, with auction web sites such as eBay, you may be able to sell your not-so-energy-efficient equipment now instead of waiting for it to break. Depending on when you bought your equipment, upgrading may or may not make sense for you right now. Eventually you will have to buy a new computer, though, and when that time comes, these are the things you should know.

What You Should Know

- Laptops are 90 percent more efficient than desktop computers because they *have* to be. Since laptops need to run on a battery for as long as possible, manufacturers build them with energy-saving features such as automatic sleep modes and low-energy, liquid crystal display (LCD) screens.

- You don't have to sacrifice quality for energy's sake. You'll find Energy Star labels on some of the best and fastest office products out there today. Plus, you're guaranteed to benefit from a lower operating cost.

- Inkjet printers are 90 percent more energy-efficient than laser printers.

- *Climate Results:* If every household that owned a photo-copier used the double-sided feature for just half their copying needs, they would prevent 20,000 *tons* of carbon dioxide from being emitted each year.

- *Money Matters:* An Energy Star fax machine, with its low-power sleep mode, can reduce your electricity cost of faxing by 50 percent.

Easy Ways You Can Help

- *Look for the Energy Star logo.* When you see the Energy Star logo on a piece of office equipment, it means that it has the ability to either go into a low-wattage sleep mode or to shut down completely after a certain period of inactivity. This not only saves electricity but can also lower your air-conditioning bills, since less heat is produced when the equipment is sleeping. There is a different Energy Star requirement for every piece of office equipment. For example, in order to be awarded the Energy Star label, a computer must go into a sleep mode of 15 watts or less after 30 minutes of inactivity, whereas a printer that can print zero to 10 pages a minute must sleep in 10 watts or less after 5 minutes of inactivity.

- ***Know that some Energy Stars are better than others.*** Having the Energy Star logo means only that the product has met Energy Star's *minimum* efficiency requirements. However, some products are extraefficient, so you should always look for the most efficient Energy Star product available. Also, keep in mind that the Energy Star logo does not necessarily mean that the sleep function will be enabled when you get your new product, or that your product is energy-efficient when it is *not* in sleep mode. Before you buy it, be sure to ask how to enable the sleep mode and how many watts the machine uses when it's active. Last, look for products that are built with recycled materials and that have the capability to add memory later if needed.

- ***Go for the multifunction devices.*** Energy Star defines a multifunction device as a machine whose primary function is to copy but which can also print, fax, or scan. For example, most of today's home fax machines are also copiers. These integrated machines almost always save money and energy because they use fewer total watts than separate machines would.

- ***Say no to the fast but greedy laser printers.*** Laser printers are popular because of their high quality and speed, but most people don't realize how much energy they use. A laser printer consumes 17 times as much energy as an inkjet printer, and more than three times as much energy as a desktop computer and monitor. What's worse, a typical laser printer uses one-third of its full power when it's simply on standby. The newest laser printers have much lower standby ratings, but these advances were made only recently. Check your owner's manual to see what your laser printer's standby wattage is—you may find that it's time to upgrade to . . .

- ***Say yes to the sensible inkjets.*** Inkjet printers cost about 80 percent less than laser printers, and although they are slightly slower, the newest inkjets offer the *same* print quality as laser printers. Inkjets also print better on scrap paper (for printing drafts), whereas most laser printers will jam with scrap paper. If you absolutely *must* have a laser printer, know that a slower one will use less energy, and be sure to get one that can print double-sided (duplex).

- ***Choose your fax machine wisely.*** Fax machines don't use a lot of energy while faxing, but if you need to leave yours on all day to receive faxes, it will. For this reason, the standby power wattage is much more important than the active power wattage when you're choosing a new fax machine. There are five main types of fax machines. We recommend a high-quality inkjet. It's the best in terms of quality, money, and energy use.

  →***Inkjet*** fax machines are the most energy-efficient. They have relatively high print quality, use plain paper, and are moderately priced between thermal and laser fax machines. Their only drawback is that they are slightly slower than laser fax machines.

→*Direct Thermal* fax machines are the cheapest, and they are moderately energy-efficient. However, their thermal paper is expensive, is curly, is hard to write on, fades with time, and isn't recyclable. In the long run, you'll end up spending much more money buying the expensive paper and then photocopying your faxes onto noncurly paper.

→*Thermal transfer* fax machines are almost as cheap as direct thermal and use plain paper (which *is* recyclable). However, the print quality is only average and you'll have to change the ink film often. They're also slow and use a medium amount of energy.

→*LED (light-emitting diode)* fax machines use a technology similar to that of laser fax machines. They have all the benefits of laser fax machines but are slightly less expensive and use slightly less energy.

→*Laser* fax machines have the best print quality and use plain paper, but they are also the most expensive and use the most energy.

- *Even better than an inkjet fax machine* is an internal fax modem in your computer. This allows you to send and receive faxes electronically, so you don't have to print them out if you don't want to. You'll save paper and the energy needed to run a fax machine. A phone-combination fax is also very efficient because it uses only slightly more energy than a normal phone.

- **Recycle or donate your old equipment.** It takes 12 times the amount of energy your computer uses in one year to manufacture it. Furthermore, when a cathode ray tube computer monitor is crushed in a landfill, it releases four to eight pounds of poisonous lead into the environment. The solution? If your equipment is still in working condition, donate it to a local school, library, or charity. If not, call your local government office and ask where you can recycle your old electronics. Or check out the National Recycling Coalition's web site to find an electronics recycler near you: www.nrc-recycle. org/resources/electronics/search/getlisting.php.

Investment   *More Savings, Less Carbon Dioxide*

Annual amount of *money* **saved** as a result of buying a new Energy Star laptop *instead* of a new but inefficient desktop and monitor computer:	Annual amount of CO_2 **not** *emitted* as a result of buying a new Energy Star laptop *instead* of a new but ineffecient desktop & monitor computer:	Total amount of *money* **saved** over the life of the computer (7 yrs.) as a result of buying a new Energy Star laptop *instead* of a new but inefficient desktop and monitor computer.
$25	**495 lbs.**	**$175**

Assumptions: Both new computers cost the same (around $1,300). The nonsleeping PC and monitor use a total of 322 kilowatt-hours per year, the sleeping laptop uses 21 kilowatt-hours per year. Both are on or sleeping for 8 hours per day, five days a week, 52 weeks a year, at 8.16 cents per kilowatt-hour.

Search For More Info

- http://yosemite1.epa.gov/estar/consumers.nsf/content/home office.htm—Check out this Internet site to learn all about Energy Star computers, printers, et cetera, and to find a store near you that sells them. You can also look up the product name and number on the Energy Star lists of qualified products and then go to the company's web site to see if you can buy it on-line.

- www.recyclingadvocates.org/wepsi/ps.htm#problem—Go here for an explanation of why electronics in particular are clogging up our landfills.

- www.nrc-recycle.org/resources/electronics/managing.htm— This brief article explains how and why you should donate, sell, or recycle your old home office equipment.

Upgrade Your Water Heater

A typical 10-year-old water heater is only 50 percent efficient. Meanwhile, the newest gas water heaters on the market have efficiencies above 85 percent.

Overview. We know, we know. The only thing you really care about when it comes to your water heater is how many showers you and your family members can take before the hot water runs out. Once you realize that heating water accounts for approximately 20 percent of your monthly energy bill, however, you may want to give your water heater a little more thought. If your water heater is seven years old or older, it might make sense for you to upgrade to a new, efficient model, in terms of both your wallet and global warming.

What You Should Know

- Older water heaters are inefficient because they have to keep a huge tank of water hot at all times, ready to use at a moment's notice. This accounts for up to 30 percent of your monthly hot-water bill!

- The Environmental Protection Agency has declared geothermal heat pumps to be the most energy-efficient and cost-effective water heaters available today.

- *Climate Results:* If 500,000 households installed geothermal heat pumps for their heating, cooling, and water heating this year, they would prevent three *million tons* of carbon dioxide from being emitted—that's the equivalent of planting 190 *million* trees.

- *Money Matters:* A brand-new heat pump could cut your water-heating bill by 50 percent. For the average U.S. household, that's a savings of $100 per year. If you use the heat pump to heat and cool your home in addition to heating your water, you could save more than $400 a year on annual energy bills.

Easy Ways You Can Help

- *Know your options.* There are five main types of water heaters. The one that is best for you will depend on a variety of factors, including your climate, the size of your household, the type of fuel available, and your budget.

  1. *Solar.* Solar water heaters are the wave of the future. Once installed, they won't cost you a thing (since sunshine is free for the taking). Plus, they work in any climate. See Tip 17 for more information.
  2. *Geothermal Heat Pump.* Also called ground-source or GeoExchange heat pumps, geothermal heat pumps are

the most efficient and environmentally friendly water heaters available next to solar water heaters, emitting up to 60 percent less carbon dioxide than traditional water heaters. Geothermal heat pumps take heat from below the earth's surface and transfer it to water storage tanks, where the heat is then used to warm up water. This process works because the temperature five feet below the Earth's surface stays constant, even in the middle of winter in Wisconsin.

Geothermal heat pumps function well in both warm and cold climates and are often used to heat and cool homes in addition to water. You can install a geothermal heat pump system on nearly any size lot, and they have lower maintenance costs and longer life spans than other water-heating technologies. Last, although geothermal heat pumps cost more up-front than traditional water heaters, they'll pay for themselves within three to five years thanks to water-heating bills being reduced by up to 40 percent.

3. *Air-Source Heat Pump.* An air-source heat pump water heater works like a geothermal heat pump except that it uses warm *air*—whether from the outdoors, your furnace room, or the back of your air conditioner—to heat water instead of using heat from the Earth. An air-source heat pump water heater is a great option if you live in a climate where the outside temperature doesn't drop below 40°F. These water heaters have a high initial cost but will pay for themselves within a few years (and then start saving you money) thanks to significantly lower operating costs.

4. ***Tankless.*** Also called instantaneous, demand, or point-of-use water heaters, these devices heat water on demand, only when you need it. As a result, they are very small and do not require storage tanks. Even though tankless water heaters cost more initially than traditional water heaters, because they don't have to keep a tank full of water hot all the time, they're much more efficient (assuming you get one with a pilotless ignition—see "Pick Your Fuel" for more info). These water heaters also *never* run out of hot water, although the catch is that the hot water flows out at a limited rate. For that reason, tankless water heaters make the most sense for one- to two-person households, remote bathrooms, or vacation homes.

5. ***Storage Tank.*** Storage tank water heaters are the most popular kind of water heater in the United States. Unfortunately, they're also often the most inefficient, since they keep up to 80 gallons of water shower-hot all day, every day. However, the latest natural gas, sealed-combustion storage water heaters are up to 35 percent more efficient than older models.

- Follow these five steps.

1. ***Pick your fuel.*** Solar water heaters are fueled by sunlight, heat pumps run on electricity, and tankless and storage tank water heaters can be fueled by electricity, natural gas, propane, or heating oil. The type of fuel that's available and the costs of different fuels in your area will determine the types of water heaters you can

buy. When doing your homework, know that having a storage tank to keep hot at all times wastes about the same amount of energy as having a pilot light burning at all times. Therefore, if you're buying a natural gas water heater, be sure to get one with a pilotless ignition.

2. ***Figure out your FHR.*** To determine the size of water heater you need, you first have to calculate your household's first hour rating (FHR), that is, the amount of hot water your family uses during its busiest hour (for instance, three people showering within one hour). Then, look on any new water heater's EnergyGuide label to compare its FHR. You'll notice that a bigger water heater does not necessarily mean a higher FHR.

3. ***Compare efficiencies.*** Once you've determined what type of fuel to use and your family's FHR, you can decide which of the five systems makes the most sense for your household. Then, buy the model with the highest efficiency rating (called the energy factor or EF, which you'll also find on the EnergyGuide label). The higher the energy factor, the less the water heater will cost to operate each month. Be careful, though—you can't compare the energy factors of different fuels. Choose your fuel first and *then* compare energy factors.

4. ***Get the right bells and whistles.*** Choose a water heater that has a vacation setting, so you can easily turn it off when you're away for three or more days. If you're buying a natural gas, oil, or propane-fired water heater, make sure it has sealed combustion. If you're buying a

tankless water heater, make sure it has modulating temperature control. Last, buy a water heater that is heavily insulated so you don't have to buy a separate insulating jacket.

5. ***Look at the Life-cycle cost.*** We've said it once and we'll say it again: Whenever you're buying something that runs on a fuel (such as electricity), you need to take into account its monthly operating cost over its expected life span as well as its initial price. The cheapest water heater will most likely be the one that uses the most energy every month and breaks down after only a few years. Also, be sure to check if your utility or state offers any rebates on high-efficiency water heaters.

Investment — ***More Savings, Less Carbon Dioxide***

Annual amount of *money* **saved** as a result of installing a geothermal heat pump water heater, taking into account the initial cost of installation:	Annual amount of CO_2 **not** *emitted* as a result of installing a geothermal heat pump water heater:	Total amount of *money saved* and CO_2 **not** *emitted* as a result of installing a geothermal heat pump water heater over its lifetime (20 yrs.), taking into account the initial cost of installation:
$30	**1,860 lbs.**	**$600** **37,200 lbs.**

Assumptions: Geothermal heat pump water heater reduces water heating energy usage by 40 percent; U.S. national average household water heating costs $200 a year using 2,835 kilowatt-hours per year; system costs $1,000 to install and lasts 20 years.

Search For More Info

- http://yosemite1.epa.gov/estar/consumers.nsf/content/ghp.htm—Go here to find information on Energy Star-labeled geothermal heat pumps.

- www.blueridgeemc.com/products_services/resi_water-heaters_sizing.htm—This site will help you determine your household's first hour rating (FHR).

- www.eren.doe.gov/consumerinfo/refbriefs/bc1.html—Check out this Department of Energy web site to learn all about tankless water heaters.

Switch to Solar

Compared with an electric water heater, a solar water heater will reduce your water heating costs by up to 85 percent and cut your carbon dioxide emissions by 100 percent!

Overview. More than 80 percent of the energy consumed by Americans comes from the burning of fossil fuels, even though we have the technology to produce 100 percent of our energy using renewable sources, such as wind, hydroelectricity, or the sun. We know that residential windmills are not very prevalent yet, but solar panels are. The latest in solar technology can save you a significant amount of money while providing your entire home with heat, hot water, and electricity. Plus, sunlight is absolutely free, it won't run out for another quadrillion years or so, and it doesn't pollute—*at all.* The energy of the future truly is solar power.

What You Should Know

- There is *more* than enough sunlight to provide the entire human population with all the electricity, hot water, and heating it needs.

- As our reserves of fossil fuels shrink, the prices of these fuels—including electricity—will rise. Solar energy, by contrast, will only get less expensive as its technology continues to improve.

- Even though the initial cost of a solar water heater or photovoltaic system is high, the fuel it will use for the next 20 to 40 years is absolutely free—because it's sunshine! That means that after it pays for itself (in approximately four to eight years), you'll get hot water for free for as long as the heater lasts. Rising fuel prices won't affect you, you'll increase the resale value of your home, and best of all, you won't be polluting the environment.

- A rooftop solar energy collector has four components: First, the collector cover, usually made of glass, lets sunlight through to the absorbers. Then the absorber, a dark, flat surface, soaks up the sun's heat. The absorber can be filled with air, water, or another fluid. An insulation material, such as the glass of the cover, keeps the heat from escaping. Finally, vents or pipes carry the heated air, water, or fluid to your home.

- *Climate Results:* Just one 1,000-watt photovoltaic solar panel that converts sunlight into electricity will prevent 1.8 *tons* of carbon dioxide from being emitted each year.

- *Money Matters:* For outdoor swimming pools in all climates, solar pool heaters save an average of $700 a year compared with conventional heaters, and they can save up to $1,500 in warmer climates. They can save even more than that for indoor pools.

Easy Ways You Can Help

- *Get solar electricity!* The sun's energy can be transformed into electricity in two ways: The sun's *light* can be transformed into electricity through photovoltaic cells, or the sun's *heat* can be transformed into electricity via solar-thermal technology. We recommend using photovoltaic technology because the panels last for decades and because it creates electricity directly, without any moving parts or turbines (which means less maintenance). Furthermore, since photovoltaic cells rely on the sun's light and not its heat, they can produce electricity no matter how cold it gets, as long as the sky is somewhat clear.

  Photovoltaic cells can be used to fully power homes, RV's, and boats, or they can supplement your utility's grid power (so you're guaranteed always to have electricity). Plus, many utilities will buy any excess power your photovoltaic cells generate and give you a credit on your next electric bill! Photovoltaic panels for homes are currently quite expensive, although depending on the state you live in, you may be eligible for enough rebates and tax incentives to make them worth the investment. A qualified contractor should install these flat, dark tiles onto your roof.

- *Get a solar water heater.* A contractor can also install a solar water heater, which uses the sun's energy to heat

water for your house, on your roof. The heated water either
is pumped electrically to a tank inside your house (using
electricity from the grid or from a small photovoltaic solar
panel) or rises naturally to a tank on your roof. Solar water
heaters almost always use a conventional water heater as a
backup so you're guaranteed never to run out of hot water.
Best of all, solar water heaters will reduce your annual
water heating bill by 50 to 85 percent. Solar water heaters
can be used in any climate, although some climates will see
bigger savings than others.

- *Get a solar pool heater.* A solar water heater can easily provide up to 100 percent of your pool's heating needs, since
  pool water doesn't need to be as hot as bathwater does.
  When combined with a solar cover at night (see Tip 13),
  solar heating allows pools to be used up to an extra month
  at both the beginning and the end of summer! This
  explains why more people today are using solar technology
  to heat their pools than for any other purpose. Here's how
  it works: Once the pool water goes through the filter, it gets
  sent through solar energy collectors, which heat up the
  water. The heated filtered water then goes back to the pool.

  Since your pool already has a filter, adding the solar collectors is easy. You can keep your old water heater and use
  it as a backup, but chances are you won't need it. The only
  drawback is space, although putting the solar collectors on
  your roof will solve that problem. Last, even though the fil-

ter pump will still use a small amount of electricity, you can be completely emission-free if you hook it up to a small photovoltaic solar panel. A solar heating system will reduce your pool heating costs by an average of 40 percent, and up to 85 percent in some climates. A system will pay for itself within five to eight years.

- *Get more solar stuff.* There are many other ways to harness the sun's energy, such as portable solar ovens (great for camping!), solar attic fans (see Tip 24), and solar-powered outdoor lighting. Solar-powered outdoor lights have miniphotovoltaic cells that soak up sunlight during the day, allowing them to store enough electricity to power the lightbulbs for eight to ten hours at night.

- *Get it for free!* Many state and local governments offer tempting tax incentives, such as $1,500 income tax credits, to promote the use of solar water heaters and photovoltaic systems. Similarly, many utilities offer rebates and subsidies from $150 to $4,000, as well as no-interest loans, when you buy solar products. Why would the utility companies want you to stop paying for their electricity and hot water, you ask? Because by using solar energy, you're helping them to reduce their peak demand (during the morning and early evening). Also, if you install a photovoltaic system, keep in mind that, depending on your state, you may be able to sell any excess power that your photovoltaic system generates *back* to the utility company!

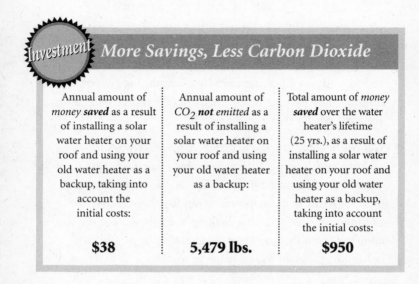

Investment — **More Savings, Less Carbon Dioxide**

Annual amount of *money* **saved** as a result of installing a solar water heater on your roof and using your old water heater as a backup, taking into account the initial costs:	Annual amount of CO_2 **not** *emitted* as a result of installing a solar water heater on your roof and using your old water heater as a backup:	Total amount of *money* **saved** over the water heater's lifetime (25 yrs.), as a result of installing a solar water heater on your roof and using your old water heater as a backup, taking into account the initial costs:
$38	**5,479 lbs.**	**$950**

Assumptions: Based on average U.S. household consuming 19 million Btus and spending $196 a year on water heating; solar water heater provides 60 percent of annual hot water, costs $2,000 including installation, and lasts 25 years.

Search For More Info

- www.bpsolar.com/contentpage.cfm?page=67—Go to this web site to learn more about photovoltaic panels for your roof.

- www.rerc-vt.org/solarbasics.htm—Here are some easy-to-understand answers to the most frequently asked questions about solar water heaters.

- www.eren.doe.gov/rspec/software.html—This free software developed by the Department of Energy will tell you exactly how much a solar pool heating system could save *your* pool.

- *Real Goods Solar Living Source Book*, by John Schaeffer. Check out this in-depth description of the very latest in solar technologies for the home.

Home Heating and Cooling

Fun with Furnaces

Nearly 50 percent of a household's annual energy costs go toward heating and cooling.

Overview. You spend the most money and use the most energy on heating and cooling your home—more than $500 a year if you're like most Americans. But that also means you have the most room to save energy and money in this area! That's why these next few tips are dedicated to different aspects of your heating and cooling systems. Let's start with the one you probably haven't looked at since it was installed: your furnace or boiler.

What You Should Know

- Most people don't know this, but the average home heating system in America was built to be only 60 percent efficient, meaning that only 60 percent of the natural gas or electricity consumed gets converted into useful heat. And if you don't properly maintain your system—as many homeowners today don't—your efficiency could be less than *50 percent*! Today's newest furnaces or boilers are built to be 95 percent efficient. If you can't afford to upgrade right now, though, the best thing you can do is to maintain your system so that it operates at its peak efficiency.

- Furnaces and boilers use more energy when they are first firing up and less when they are running steadily. If your furnace or boiler seems to start and stop a lot, get it checked out—it might be too big for your house. The fact that furnaces are more efficient when they're running steadily, however, doesn't mean you should never turn down your furnace! Turning your thermostat down for four hours or more and then turning it back on is *always* more efficient than leaving it on a high setting for that amount of time (see Tip 19 for more info).

- A simple tune-up of your heating system every one to two years can reduce your annual heating costs by up to 10 percent.

- *Climate Results:* If just half of the American households who heated their homes got a tune-up on their heating systems this year, they would prevent 30 *million tons* of carbon dioxide from being emitted into the atmosphere.

- *Money Matters:* If you change your furnace's filter once a month during heavy use and once every other month during light usage, you'll save at least $12 a year, even after taking into account the cost of the filters.

Easy Ways You Can Help

- *Va-va-va-vacuum.* When you vacuum, be sure to vacuum the floor heating vents or radiators as well—the less dust that's in your heating system, the better it will work. Also, make sure that furniture or drapes don't block the vents or radiators.

- *Close vents to unused rooms.* If you won't be using a room for two or more days, shut the door, close the heating vents (or turn off the radiators) in that room, and turn down the thermostat for that room (if you have zone heating)—this will reduce your heating costs by up to 10 percent! Be careful not to close too many heating vents in your home, though— that will put stress on your furnace's fan.

- *Change your filters!* A dirty, dusty furnace filter reduces airflow, making your system less efficient. Every month that your furnace is being used heavily, replace its air filter (if you have disposable ones) or clean it (if you have a permanent one). You can pick furnace filters up at your local hardware store for about a dollar each. If you own dogs, cats, or other furry pets, you may need to change your filter more than once a month. Replace or clean the filter every two to four months when you're not using your furnace heavily.

- *Get to know your furnace fan.* Your furnace's fan is what blows the hot air created by the furnace to the rest of your house. If it is set to turn on at too high a temperature, it will turn on later than it should. Check to make sure it's set to turn on between 100° and 110°F (if there are two settings, have it come on between 100° and 110°F and turn off between 80° and 90°F).

- *Just for boilers:* Go to the hardware store, pick up some radiator reflectors, and put them behind any radiators that are next to exterior walls (walls that face the outside). This will help prevent heat being lost through the wall. Also, once every winter, hold a bowl under your radiator's valve and slowly open the valve to let out any unwanted pockets of air. Once the water is flowing steadily, close the valve back up.

- *Get a tune-up!* Tuning up your heating system is the best thing you can do to improve its efficiency and cut your energy costs (next to upgrading to a superefficient model— see Tip 28). A $50 to $100 tune-up will also extend your furnace's life span. A system that runs on oil should be tuned up once a year, a system that runs on natural gas should be checked every two years, and an electric heat pump or resistance heater should be tuned up every two to three years. The company that sells you your heating fuel or electricity most likely has a technician who can perform these tune-ups.

Low Cost

More Savings, Less Carbon Dioxide

Annual amount of *money* ***saved*** as a result of getting a $50 tune-up on your heating system every two years, taking into account the cost of the tune-up:	Annual amount of CO_2 ***not*** *emitted* as a result of getting a $50 tune-up on your heating system every two years:
$17	**1,248 lbs.**

Assumptions: Based on average U.S. household that uses 52 million Btus and spends $421 per year on heating; tune-up reduces heat usage by 10 percent; 100,000 Btus = 1 therm; 1 therm emits 12 pounds carbon dioxide.

Search For More Info

- www.acdoctor.com/heating/heating.htm—Here is an easy-to-understand explanation of how your heating system works. Also, check out this glossary of terms to help you get a handle on your heating system: www.comfortnet.com/glossary/glossary.htm#TOP.

- www.commerce.state.mn.us/pages/Energy/InfoCenter/furntune.htm—Check out this web site to learn about more complex things you can do to improve your furnace or boiler's efficiency and what you should expect from a tune-up.

- www.servicemagic.com—Go here to find a prescreened heating contractor in you area. Select "Heating and Cooling" from the list of categories.

Do the Right Temp

Heating homes in America produces 310 million tons of carbon dioxide emissions a year.

Overview. Sometimes technology gets it right. Sometimes new inventions are convenient *and* good for the environment. The programmable thermostat is one of these. Also known as a clock or setback thermostat, this energy-conscious device allows you to program different temperature settings for when you're at home, at work, awake, or asleep. It also allows you to set different programs for different days (including the weekends), depending on your schedule. And the best part is that it's completely automatic. Once you program it, you never have to touch it again! Imagine waking up to a warm bedroom or coming home to a warm house and knowing that you didn't waste energy all night or day long. Why didn't we think of this sooner?

What You Should Know

- Programmable thermostats can store up to six daily temperature settings—one for when you wake up, one for when you go to work, and so on. You can also program different settings for the weekend versus the workweek. Finally, programmable thermostats automatically adjust what time they turn on as outdoor temperatures change with the seasons.

- If you want the house hotter or colder you can always temporarily override the current setting for three hours without affecting the program.

- Programmable thermostats typically replace existing thermostats, although some battery-powered ones can be placed over your current thermostat.

- Small liquid crystal display (LCD) screens, like you would find on a laptop computer, make programmable thermostats energy-efficient and easy to read. Most also have backup batteries so that your settings don't get erased in the event of a power outage.

- *Climate Results:* Approximately 44 percent of American households own programmable thermostats. Seventy-three percent of these households, however, don't use them—they use only the manual controls! If this 73 percent started using the programmable feature on their thermostats, they could prevent 15 *million tons* of carbon dioxide from being emitted each year.

- *Money Matters:* Programmable thermostats pay for themselves in approximately one year thanks to annual energy savings of up to 30 percent.

Easy Ways You Can Help

- *How low can you go?* Turning down your thermostat is the best way to save money and energy on your heating bill, but we don't expect you to freeze! Turning your furnace down even 1°F while you're home can save a significant amount of energy, and you probably won't even notice a difference in comfort. Turning it down 10°F while you're gone during the day and sleeping at night can reduce your annual heating bill by approximately 15 percent! The same principles are true during the summer when it comes to your air conditioner: Turning your air conditioner's thermostat up just 1°F can reduce your cooling bill by 2 percent, and turning it up by 10°F while you're gone during the day can reduce your cooling bill by as much as 20 percent!

  The Department of Energy recommends keeping your thermostat set at 68°F (or lower) during the winter when you're home, and 60°F while you're gone during the day or sleeping. For the summer, it recommends setting your thermostat to 78°F (or higher) while you're home, and 85°F while you're at work. Last, if you'll be away from home for more than three days in the winter, turn your thermostat down to 55°F—this is still warm enough to keep your pipes from freezing. If you'll be gone for more than a day during the summer, turn your air conditioner completely off.

- *Don't crank it.* In the winter, turning your thermostat up to 82°F to warm the house quickly won't heat it up any faster than turning the thermostat to 72°F will. Also, you probably won't remember to turn it back down until you're burning up, which will waste even more energy. So, whether it's summer or winter, don't crank your thermostat.

- *Steady as she goes.* Try to keep the temperature of your house fairly constant for stretches of four hours or more. Frequently adjusting your thermostat can cause your furnace to turn on and off needlessly, wasting energy. Make significant thermostat adjustments when you leave for the day or go to sleep, but otherwise, try not to change your thermostat.

- *Go programmable.* Programmable thermostats are the best way to go if you want easy-to-achieve energy savings that let you continue to live in comfort. Once it's programmed, you won't have to touch your thermostat until spring! When choosing a programmable thermostat, look for the Energy Star logo, which means that the thermostat has at least two programming periods (weekdays and weekend) and the ability to program at least four temperature settings per day (awake, daytime, evening, and asleep). (See Tip 14 for more info on Energy Star products.)

- *Location matters.* A programmable thermostat won't save you any money if you place it next to a hot lamp or a cold window. Thermostats are extremely sensitive, so you want to make sure they're reading your house's temperature accurately. Put some insulation in the hole behind the thermostat, and don't place any lamps, TV's, or other heat-emitting appliances too close to it. Also, make sure it's in neither a drafty nor a sunny part of your house.

More Savings, Less Carbon Dioxide

Annual amount of *money **saved*** as a result of turning your thermostat down (either manually or with a programmable thermostat) 10°F while you're sleeping at night and 1°F during the day, taking into account the initial cost of the thermostat:

$46

Annual amount of CO_2 ***not** emitted* as a result of turning your thermostat down (either manually or with a programmable thermostat) 10°F while you're sleeping at night and 1°F during the day:

811 lbs.

NOTE: These savings will also apply to your summer cooling bill if you turn the thermostat of your air conditioner up 10°F while sleeping and up 1°F the rest of the day.

Assumptions: Temperature is lowered 10°F for 7 hours while sleeping, which reduces heating bill by at least 10 percent; temperature is turned down 1°F for the other 17 hours, which reduces heating bill by at least 3 percent; based on national average U.S. household spending $421 per year and using 52 million Btus per year on space heating; programmable thermostat costs $60 and lasts at least seven years; 100,000 Btus = 1 therm; 1 therm emits 12 pounds carbon dioxide.

Search For More Info

- www.eren.doe.gov/erec/factsheets/thermo.html—Go here to learn about the five types of programmable thermostats and to decide which one is right for you.

- http://yosemite1.epa.gov/estar/consumers.nsf/content/prgth erm.htm—Check out Energy Star's Internet site on programmable thermostats, including information on where to buy one near you.

- www.truevalue.com—Go here and type "programmable thermostat" in the search box to buy programmable thermostats on-line.

Assess the A/C

For every room that you close off and don't air-condition, you'll reduce your annual air-conditioning costs and energy usage by approximately 10 percent.

Overview. Nobody likes to sweat. Knowing that air-conditioning makes up at least 12 percent of your annual electricity bill, however, can sure make you want to loosen your collar a bit! Air-conditioning is a luxury that carries a high price—not only on your electric bill but also on the environment. Americans emit 100 *million tons* of carbon dioxide every year running their air conditioners. Luckily, there are ways to make sure your air conditioner is running at peak efficiency.

What You Should Know

- Seventy-two percent of Americans use air-conditioning to cool their homes. Altogether they spend $10.2 billion a year on air-conditioning alone.

- A typical American household spends $175 a year on air-conditioning. There are two basic types of air conditioners: room air conditioners that fit into a window and central air conditioners that cool an entire house. As you might expect, room air conditioners are less expensive than central systems because they cool only the rooms in which they are placed. However, central air-conditioning is more efficient than multiple room air conditioners if you're trying to cool an entire house.

- **Climate Results:** If every American who used air-conditioning turned up his or her thermostat by just 1°F this summer, they would prevent 5.5 *million tons* of carbon dioxide from being emitted into the atmosphere.

- **Money Matters:** If you replace your air conditioner's disposable filter when it gets dirty (or rinse your permanent filter with water), you'll save approximately $15 a year, and that includes the cost of the new filters.

Easy Ways You Can Help

- **Don't cool an empty room.** Close the doors and vents to rooms that aren't being used, or turn off the room air conditioners in those rooms.

- **Clean those coils.** Your air conditioner has two coils: a cold, indoor one called the evaporator coil and a hot, outdoor one called the condenser coil. Both are covered by small aluminum fins. When the coils get dirty, they lose their ability to absorb or expel heat. The inside coil collects dust easily because it is constantly wet from the air conditioner. The outside one collects dirt simply because it's outside. Every spring, if they're dirty, have a certified service person come to clean the coils. If the fins covering the coils appear dirty, you can vacuum or wipe them off yourself. If the fins are bent, straighten them with a plastic spatula so they don't restrict airflow.

- *Check the refrigerant.* Once every two years, or if your air conditioner doesn't seem to be working well, have a trained service person check the level of the refrigerant (the fluid that makes your air conditioner work). If it is even slightly above or below the manufacturer's specified level, your air conditioner will not perform efficiently. If it is consistently low, you could have a leak, which the service person should fix immediately. Also, while the service person is there, have him or her check the compressor, fan, and electrical connections to make sure everything is working properly.

- *Set it right.* If your air conditioner has one, always select the recirculate setting instead of the outside air setting. The recirculate setting will recool the inside air instead of cooling new, hot air from outside. Also, always set your air conditioner's fan to its highest speed, except on very humid days. A high fan speed saves energy because it blows the cooled air into your house more quickly (a low fan speed is better at removing humidity, but on nonhumid days it's not necessary). Last, try to use the automatic setting as opposed to the continuous fan/ventilate/on setting. The continuous fan setting will run your air conditioner's fan constantly, even when the coils aren't actually cooling any air. If you like the ventilation it provides, keep your air conditioner on automatic and run ceiling fans instead. They use much less electricity and therefore cost less to operate (see Tip 21).

- *Clean those filters.* Replacing your air conditioner's filter when it gets dirty is the easiest way you can improve its efficiency. A dirty filter will restrict airflow, making your air conditioner slower and less effective. Check the filter by holding it up to a light—if you can't see through it, you should replace it or clean it. You can pick up disposable filters at your local hardware store for about a dollar each. Reusable (washable) filters are only slightly more expensive, but they're more environmentally friendly and they last longer.

Especially for Room Air Conditioners

- *Seal it.* Every spring, check the seal between your room air conditioner and its window frame. If the seal is not airtight, put weather stripping around the window to prevent air-conditioned air from escaping to the outside (see Tip 23 for more info).

- *Unclog it.* If the drain channels in your room air conditioner are clogged, they won't be able to remove humidity. Push a stiff wire through these channels to unclog them.

- *Store it.* Since room air conditioners tend to be drafty no matter how well you seal them to your windows, it's a good idea to remove or at least cover them during the winter.

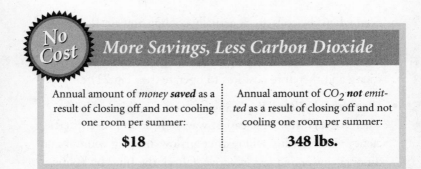

No Cost

More Savings, Less Carbon Dioxide

Annual amount of *money* **saved** as a result of closing off and not cooling one room per summer:

$18

Annual amount of CO_2 **not** emitted as a result of closing off and not cooling one room per summer:

348 lbs.

Assumptions: Based on average U.S. household central air system that uses 2,123 kilowatt-hours per year, costs $173 a year, and cools 1,823 square feet; electricity costs 8.16 cents per kilowatt-hour; closing off one room reduces cooling use by 10 percent.

Search For More Info

- www.howstuffworks.com/ac.htm—Check out this easy-to-understand, illustrated explanation of how air conditioners work.

- www.eren.doe.gov/erec/factsheets/aircond.html—This in-depth fact sheet written by the Department of Energy explains how to make your air conditioner as efficient as possible.

- www.eren.doe.gov/consumerinfo/refbriefs/bd6.html—This brief article explains what chlorofluorocarbons (CFC's) are, what products they're in, and what's being done about them.

Alternative Ways to Cool

Using ceiling fans instead of an air conditioner will reduce your cooling costs by at least 60 percent.

Overview. Air conditioners weren't always around. Before they were, people kept cool in other, more energy-efficient ways. It's not always a good idea to return to the way things were, but sometimes it *is* a good idea to go back to the basics—especially when those basics are environmentally friendly, inexpensive, and convenient! Turning on a ceiling fan is no harder than turning on your air conditioner, but it *is* a lot cheaper. The more we use inefficient, energy-guzzling air conditioners, the sooner we'll be living on a planet where air-conditioning will be necessary 365 days a year. Keep reading to learn how you can stay just as cool *without* having to turn on the A/C.

What You Should Know

- Humidity makes us feel hotter than it really is. That's because our bodies sweat to stay cool, and sweating is a type of evaporation. When it's humid out, it's harder for water to evaporate, which means it's harder for us to sweat and stay cool. That's why air conditioners remove both heat *and* humidity from the air. In humid climates, a dehumidifier can even replace an air conditioner.

- A breeze makes us feel cooler than it really is, because wind makes it easier for water to evaporate and therefore easier for us to sweat and stay cool. The human comfort zone during the summer is 72° to 78°F, but with the breeze from a ceiling fan making us *feel* cooler, that comfort zone can be raised to 78° to 82°F.

- Ceiling fans used in conjunction with your air conditioner will help to spread cooled air through your house. In the winter, you can reverse the direction of your ceiling fans and put them on a low speed to help evenly distribute warm air that collects near your ceiling.

- Evaporative coolers, also known as swamp coolers, pull hot, dry outside air over wet pads inside the coolers. The drier outside air absorbs some of the water from these pads and becomes cooler as a result. A fan then blows the now cold air into your house. Because their technology is so much simpler than that of air conditioners, swamp coolers use about one-fourth as much energy.

- *Climate Results:* Forty-one percent of Americans who use air-conditioning use it all summer long. If only one-fourth of them used a dehumidifier instead of their air-conditioning for only one-fourth of the summer (at the beginning and end), they would prevent three *million tons* of carbon dioxide from being emitted each year.

- *Money Matters:* Evaporative coolers use 75 percent less energy than air conditioners. If you used an evaporative cooler instead of an air conditioner for just half the summer, you'd save at least $65 a year. If you used one for the whole summer, you'd save at least $130 a year.

Easy Ways You Can Help

- *Use ceiling fans.* In milder climates, ceiling fans can actually replace air conditioners. In hotter climates, they can be used in conjunction with an air conditioner, allowing you to turn up the thermostat by 6°F while still feeling comfortable. This substitution can have a dramatic impact on your cooling costs, since running a ceiling fan for 24 hours on high speed costs only about 35 cents. Ceiling or window fans are also good options for cooling individual rooms.

  Open your windows at night and turn on ceiling fans in order to ventilate your house instead of using the fan-only or ventilate switch on your air conditioner (which consumes more electricity than a few ceiling fans do). Last, be sure to buy ceiling fans with the Energy Star label, which tells you that they are 20 percent more energy-efficient and can save you between $20 and $30 a year over standard ceiling fans.

- *Try an evaporative cooler.* These energy-saving devices use a natural cooling process that can lower the temperature of outside air by up to 30°F. They're also excellent ventilators—an evaporative cooler will replace an entire room full of air every one to three minutes! Because they add humidity to the air they cool, however, they aren't suitable for humid climates. Two-stage models that produce cool, *dry* air are available for these climates. Last, some utilities offer rebates of up to $300 for buying whole-house evaporative systems.

- *Consider a dehumidifier.* If you live in a humid climate, this one's for you! A typical dehumidifier uses less energy than a room or central air conditioner does, and on many days you may be able to use it *instead* of your air conditioner. Reducing humidity will make you *feel* cooler even though the actual temperature may still be high. Don't use a dehumidifier and an air conditioner at the same time, though (since an air conditioner acts as a dehumidifier anyway).

- *Button up during the day.* Keep the heat and sunlight out of your house during the day by closing all windows, doors, drapes, and blinds. Closing drapes or blinds is especially important on windows that face east and west. Try to reduce the number of times you open doors to the outside. However, if it's a mild day (cooler than 78°F), turn off the air conditioner, open all windows and doors, and enjoy the breeze.

- *Ventilate at night.* In milder climates, ventilating your house with cool night air can allow you to turn off your air conditioner altogether during the day. Opening all drapes and blinds after dark will allow the heat in your house to escape through the windows. Opening windows on opposite sides of the house and turning on some fans will circulate cool night air through your home. Don't ventilate with outside air unless it's 78°F or lower, though.

- ***Don't cook dinner . . .*** until later! Try to avoid doing things that add heat or humidity to your house during the middle of the day, such as cooking with an oven or stove or running your dishwasher. If you must use the stove, use the exhaust fan to suck the steam out of your house (but don't leave the exhaust fan on for too long—see Tip 10). Also, don't use a lot of incandescent lightbulbs, since they give off 90 percent heat—install compact fluorescent bulbs instead (see Tip 1).

- ***Add a radiant barrier.*** Another way to cool your house is to stop the sun's heat from seeping through your roof, into your attic, and down into your house. One way to do this is to add a radiant barrier—an inexpensive type of aluminum foil—on the underside of your roof. This will stop 95 percent of the heat from radiating into your attic, and you can easily install it yourself. You can also add it to your attic floor to stop heat in your house from seeping up into the attic during winter. If you buy a vented, multilayered radiant barrier, it will also act as insulation. Radiant barriers have the best results in hot climates or in homes with little insulation.

Search For More Info

- www.energystar.gov/products—Scroll down and click on Energy Star "dehumidifiers" and "ceiling fans" to learn how much they can save you and where you can buy them.

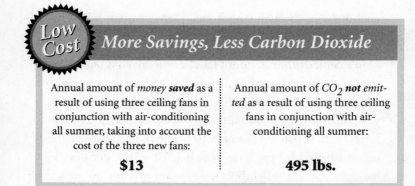

Low Cost

More Savings, Less Carbon Dioxide

Annual amount of *money **saved*** as a result of using three ceiling fans in conjunction with air-conditioning all summer, taking into account the cost of the three new fans:

$13

Annual amount of CO_2 ***not*** *emitted* as a result of using three ceiling fans in conjunction with air-conditioning all summer:

495 lbs.

Assumptions: Based on average U.S. household use of 1,677 kilowatt-hours per year and $140 per year on air-conditioning; three fans allow a 6°F rise in temperature, which reduces cooling costs by 18 percent; each fan is self-installed, costs $60, and lasts 15 years.

- www.easy2diy.com/tutorials/diy0163/index.asp—Check out this do-it-yourself guide on installing a ceiling fan. Go here, www.dulley.com/docs/f411.htm, to learn how to install a radiant barrier in your attic.

- www.consumerenergycenter.org/homeandwork/homes/inside/heatandcool/evaporative_coolers.html—Go here to learn everything you ever wanted to know about evaporative coolers.

- www.rmi.org—The Rocky Mountain Institute is a great environmental organization that provides in-depth information on global warming and other issues. Check out their "Climate" link.

Dress Up Your Windows

*Sun-control window screens can block up to
90 percent of the sun's heat.*

Overview. With the right set of curtains, you can get lower heating and cooling bills, a more comfortable home, and some great decoration, all from one piece of fabric. Shading your windows in the summer will keep your house considerably cooler, and with new sun-control shades and screens, you won't even have to give up the view. Covering your windows on cold nights will keep your house warmer in the winter, which means your furnace won't have to work as hard. If Americans paid as much attention to their window accessories as they did to their fashion accessories, we'd be a lot closer to getting a handle on global warming.

What You Should Know

- White window coverings, such as drapes, blinds, or awnings, can reduce solar heat gain in the summer by 50 percent.

- In the winter, up to 16 percent of the air your furnace works so hard to heat up is lost through uncovered windows.

- Exterior window shades, such as awnings or shutters, are 50 percent more effective at reducing solar heat gain than interior shades, such as curtains or blinds, because they block the sunlight *before* it hits your windows.

- Many rolling shutters (shutters that can roll all the way down or up) have insulation in their slats, which can improve the insulating efficiency of a single-pane window by up to 65 percent and that of a double-paned window by up to 50 percent!

- *Climate Results:* An awning can reduce a window's solar heat gain during the summer by up to 77 percent. That means if 100,000 people put awnings above their largest south-facing windows this year, they would collectively prevent more than 14,000 *tons* of carbon dioxide from being emitted.

- *Money Matters:* If you install sun-control screens on your windows, you'll save at least $45 in air-conditioning costs every summer.

Easy Ways You Can Help

- ***Add some curtains.*** Special insulating window shades or curtains can keep warm air in during the winter and prevent outside heat from passing into your home during the summer. Also, consider buying *white* curtains, blinds, or shades, since white is better at reflecting the sunlight away from your house than are darker colors, which absorb heat. Last, the newest sun-blocking, summer shades are see-through, so the sun's heat is blocked but its light isn't.

- **And know how to use them.** In the winter, open up your blinds or curtains on any east-, south-, or west-facing window during the day to let sunlight (and its heat!) in. Close them as soon as the sun goes down to retain the heat. Do the exact opposite in the summer while ventilating with a ceiling fan to keep your house cool. If you have a large window that doesn't receive a lot of sun, keep its blinds or curtains closed during the night *and day* in winter to reduce heat loss. Along these same lines, place a dark rug on the floor of your sunniest room in the winter so it will absorb the sun's heat. Replace it with a light-colored rug in the summer.

- **Install an awning.** Exterior window shading is more effective at reducing solar heat gain than interior shading (blinds or curtains), and it has the added bonus of not making your rooms totally dark. The only downside to awnings is that they can be expensive. If your awning is white or light-colored, it will be even more effective. Awnings and roof overhangs are most effective on south-facing windows. In the winter, retract or remove any awnings so you can benefit from the sun's heat.

- **Get rollin'.** Rolling shutters are another popular and economical option for exterior window shading. You can close them tight for excellent insulation, privacy, and protection from severe weather. Open them slightly for ventilation and some sunlight. Roll them up completely and they will be hidden in small, rectangular casings above your windows. Remote control makes them superconvenient.

- **Install sun-control screens.** If you like fresh air in the summer but don't like the added heat you get from having your windows open, this tip's for you. Sun-control or solar screens are just like regular window screens except that they're made of tightly woven fiberglass or polyester that has been specially designed to reduce solar heat gain. You can see through them and air can flow through them just as well as through regular screens, although they may appear darker when you are looking at them from the outside (which may be a good thing, if you like privacy). Sun-control screens are available in a variety of sizes and can be easily removed for the winter. You can install them yourself or have a professional do it. Some utilities offer rebates or other incentives to buy them.

Low Cost

More Savings, Less Carbon Dioxide

Annual amount of *money* **saved** as a result of installing white curtains on south-, east-, and west-facing windows and closing them during the hottest times of the day in summer, taking into account the initial cost of the curtains:

$20

Annual amount of CO_2 **not** emitted as a result of installing white curtains on south-, east-, and west-facing windows and closing them during the hottest times of the day in summer:

481 lbs.

Assumptions: Based on average U.S. household's use of 5.7 million Btus and $140 per year on air-conditioning; 35 percent of that use is caused by solar heat gain; white shading reduces heat gain by 50 percent; basic white curtains cost $75, installed yourself, and last 15 years.

Search For More Info

- www.dulley.com/gtopics.shtml—Click on the "Window Improvements" link, then scroll down to Bulletins 465 and 736 to read more about awnings, 693 to learn about rolling shutters, and 477 to learn about sun-control screens.

- www.consumerenergycenter.org/homeandwork/homes/inside/windows/shades.html—Check out this informative fact sheet to learn more about awnings, shutters, screens, and other window coverings.

- www.openhere.com/hag/homes/home-products/blinds— Here is a list of web sites that sell all kinds of blinds, curtains, and window coverings. On the left side of the page, there's a link to "Shutters," and at the very bottom of the page, there's a link to "Awnings." Happy shopping!

Plug Air Leaks

Adding caulking and weather stripping to your windows and doors can reduce your heating and cooling costs by at least 10 percent.

Overview. Do you ever hear the wind whistling through your house on a windy day? That's the sound of warm air from your furnace escaping to the cold outside. In the summer, the opposite happens: your precious air-conditioned air escapes to the muggy outdoors. Since we assume that you'd rather not pay to heat or air-condition the outdoors, here are some tips to help you plug all those pesky air leaks, thereby saving some energy and some cash.

What You Should Know

- Contrary to popular belief, insulation does *not* block the flow of air. Insulation stops only the flow of *heat*. If the insulation is densely packed, like loose-fill cellulose insulation, it will significantly *reduce* your air leakage, but it still won't stop it completely.

If you add up all the tiny air leaks in an average house, not including leaks from ducts, they're equivalent to leaving a three-foot-square window wide open!

- *Climate Results:* If just one-fourth of the households in America weather-stripped and caulked their windows and doors this weekend, the reduction in heating and cooling costs would prevent eight *million tons* of carbon dioxide from being emitted this year—and every year after.

- *Money Matters:* Weather-stripping a front and back door will cost around $25 and will save you about $30 a year in reduced heating and cooling costs. And since weather stripping lasts at least five years, that's a total savings of $125 for just an hour's worth of work. Not bad!

Easy Ways You Can Help

- *Check for individual leaks.* Most people know to start checking for air leaks around doors and windowpanes, but the biggest leaks are usually harder to find, such as where your attic, crawl space, or basement connects to your heated living space. Go to the following seven places and note if you can see an opening or feel even the slightest bit of air flowing in or out:

    1. Wherever a wall meets a floor, ceiling, doorframe, or another wall

2. Around the edges of light switches, electrical outlets, and light fixtures

3. Around fireplace flues, dryer vents, and range hood vents

4. Wherever pipes go through walls or ceilings: check behind the washing machine, underneath the sink, behind the shower and toilet, et cetera

5. On the outside of your house: between the siding panels, where the chimney and siding meet, where the foundation meets the house, where the cable lines go into the house, et cetera

6. Around all windows and exterior doors (including skylights and storm windows)—if you can rattle them or see light shining in around them, air is leaking through!

7. In your attic—this is the most common place to find air leaks, since warm air from your heated home has a natural tendency to rise into the attic any way it can. Check for air leaks around your attic door or pull-down stairs, around any wiring, piping, or ductwork that enters the attic through your ceilings or walls, and where your chimney goes into the attic. A dirt streak on your attic insulation means there's an air leak underneath (where dusty air has been escaping and creating a stain).

- *Burn some incense.* If you're having trouble finding air leaks, close all your windows, exterior doors, and fireplace flue. Then turn off any gas-burning appliances, such as furnace, water heater, or stove (this is for safety reasons). Now turn on all exhaust fans in the kitchen and bathrooms, which will suck the air out of your home. Your house will need to replace the air that's being sucked out, but since the doors and windows are all closed, it will have to get that air from any leaks. Now take a lit incense stick and walk around to those seven places we just listed. If the smoke wavers or blows horizontally, you've found a leak! (Note: You can also hire a contractor to find leaks in your house performing a blower door test.)

- *Seal those leaks!* Now that you've identified your leaks, tackle the largest ones first and make your way on down to the smallest of cracks. Remember, every leak counts!

  →*Use Caulk.* Caulking is good for leaks that are less than a half-inch wide. Pick some up at the hardware store, choose the appropriate type of caulk for what you're caulking (exterior, interior, windows, et cetera), and follow the directions carefully. Caulking is used to seal around almost everything: doors, windows, soffits, medicine cabinets, heating registers, light switches, and electrical outlets; pipes or cable wires that go through walls; where walls, floors, and ceilings meet; and any leaks to the outside.

  →*Use Foam Sealant.* You can pick up foam sealant or foam caulking at your hardware store and use it to seal gaps that are more than a half-inch wide. There are three main types of foam sealant (expanding, nonexpanding, and sprayed in), and which one you choose will depend on how big the leak is that you're filling and whether it's easy to reach. Don't use expanding foam on doors or windows.

→ ***Use Weather Stripping.*** Weather stripping is a strip of material that you attach around the edges of exterior doors, attic doors and hatches, and windows in order to make them airtight. It's inexpensive, and you can find it at your local hardware store. Pay close attention to the thresholds (bottom edges) of your doors—if weather stripping is not enough to plug the leak, you can attach a door shoe (a plastic flap that sweeps along your floor).

→ ***Tackle the Big Ones.*** To fix a big leak in your attic, staple a plastic sheet over it and caulk around the edges of the sheet. You can also greatly reduce (but not stop completely) air leakage in your attic by having a contractor blow high-density cellulose insulation into large or hard-to-reach spaces.

→ ***Special Considerations:*** When sealing around your fireplace flue, chimney, kitchen exhaust fan, or dryer vent, be sure to use heat-resistant, noncombustible caulk or foam. Second, make sure all electrical outlets (even the ones in your basement) have covers on them, and install rubber gaskets behind all outlets and switches to prevent air leakage. Last, if you have lighting fixtures that are recessed into the ceiling, make sure that they are airtight "Insulated Ceiling" (IC) fixtures, and then seal around them with heat-resistant caulk.

- **When you're starting fresh . . .** The best time to make your home airtight is when you're building a new house or remodeling your existing one. Ask to have a house wrap, an air barrier that is placed outside the exterior sheathing but underneath the siding, installed. Most important, keep an eye on your contractor and make sure he or she does a *thorough* caulking job!

Low Cost — More Savings, Less Carbon Dioxide

Annual amount of *money **saved*** as a result of caulking and weather-stripping your home's air leaks (not including ducts), taking into account the cost of the caulk and weather stripping:	Annual amount of CO_2 *not emitted* as a result of caulking and weather-stripping your home's air leaks (not including ducts):
$50	**692 lbs.**

Assumptions: Based on average U.S. household use of 57.7 million Btus and $561 per year on heating and cooling; caulking and weather-stripping reduce heating and cooling costs by 10 percent; caulking lasts 10 years and costs $12; two weather-stripping installations last five years each and cost $25 each; 100,000 Btus = 1 therm; 1 therm emits 12 pounds carbon dioxide.

Search For More Info

- http://doityourself.com/energy/weatherstripping.htm—This do-it-yourself guide gives detailed instructions on how to weather-strip your doors.

- www.ianr.unl.edu/pubs/consumered/heg157.htm—This web site gives step-by-step instructions on how to caulk and how to select the types of caulks you should use for different leaks.

- www.homeenergy.org/hewebsite/consumerinfo/air-leaks/index.html—Check out this in-depth article about caulking and weather-stripping the air leaks in your home.

Air Out the Attic

Ventilating your attic can reduce your air-conditioning costs by up to 10 percent.

Overview. Now that you've stopped air from flowing where it shouldn't, it's time to make sure air is flowing where it *should*—in your attic. In the summertime, your attic gets a double whammy: warm air from inside your house rises into the attic, and the sun's heat blasts down through the roof. As a result, your attic can turn into a veritable sauna. What's worse, all that built-up heat seeps into the rooms directly below your attic, making them up to 10°F warmer than they should be, which means your air conditioner has to work harder than it ought to.

Winter is a different story. In the winter, warm air from inside your house rises, but we hope your insulation stops this precious heat from escaping into the attic (see Tip 25). Insulation, however, stops only the *heat*. What it doesn't stop from getting into your attic is the air itself, which has moisture in it. If this air is not immediately vented to the outside, the moisture will condense into water vapor and soak down into your insulation. This makes your insulation less effective, which means more of your home's heat will rise into your attic. Your furnace will therefore have to work harder to keep your house warm. Confused? Don't be. There's a simple solution: Ventilate your attic.

What You Should Know

- Your attic can reach up to 150°F in midsummer.

- Not only will ventilating your attic reduce your cooling costs in the summer and your heating costs in the winter but it will also help prevent mold from forming and wood from rotting in your attic (both consequences of moisture condensation). Ventilating your attic also keeps the temperature of your roof uniform, which prevents ice dams from forming on the outside of the roof during winter. Last, ventilating your attic prevents dusty and possibly moldy attic air from seeping down into your living space. The only question is, Why *wouldn't* you want to ventilate your attic?

- Poor attic ventilation decreases the life of your roof. For example, curled-up roof shingles are telltale signs of poor attic ventilation—and of one very hot attic!

- A whole-house fan installed in your attic can reduce your home's indoor temperature by up to 10°F, making air-conditioning unnecessary in some climates or at certain times of the day (see next section for more information).

- ***Climate Results:*** If just 100,000 households in milder climates replaced their air conditioners with whole-house fans, they would prevent 121,000 *tons* of carbon dioxide from being emitted each year.

- *Money Matters:* Adding soffit and ridge vents to an unvented attic will reduce summer air-conditioning costs by 10 percent.

Easy Ways You Can Help

- *Add a vapor barrier.* Your first line of defense against moisture condensation in your attic is sealing off any air leaks that lead from your living space into the attic (see Tip 23). Your second line of defense is sealing your attic, top to bottom, with a vapor barrier. To do this, first see if the underside of your attic's floor insulation is covered with paper. Alternatively, plastic, tar paper, or kraft paper may be laid under the insulation. If so, you already have a vapor barrier. If not, don't worry, you don't have to rip up all your insulation. You can simply paint the ceilings of the rooms below your attic with a special vapor barrier paint, which you can pick up at your local hardware store. Last, make sure there is a vapor barrier between your roof and the insulation that is directly underneath your roof. This will prevent moisture from seeping into your attic from the outside, especially during a humid summer.

  Of course, some moisture will inevitably get into your attic. That's why your third and final line of defense against moisture condensation is ventilating your attic . . .

- **Add attic vents.** Vents along the bottom edges of your roof, called soffit vents, allow outside air to flow in, up, and then out of your attic through vents at the peak of your roof, called ridge vents. The result? A wind-ventilated attic that requires no electricity. For the best ventilation possible, make sure there are an equal number of soffit vents and ridge vents and distribute the soffit vents evenly around the bottom edges of your roof. You can either install these attic vents by yourself (see "Search for More Info") or hire a contractor to do the job.

- **Or get a solar attic fan.** A fan installed in your attic will ventilate just as well as a soffit-ridge vent system, but if it's electric, it will cost you around $10 a month. If it's a solar-powered fan, however, you won't have to pay a dime after it's installed. Plus, a solar fan changes its speed in direct proportion to the strength of the sun, so the hotter it is, the more strongly your solar fan will ventilate!

- **Install a whole-house attic fan.**

  →**What Is It?** A whole-house fan ventilates not only your attic but your entire house, helping to keep the house cooler. In fact, a whole-house fan can completely replace your air conditioner if you don't live in an extremely hot climate. Whole-house fans use 90 percent less electricity and cost significantly less than air conditioners, and they're more effective and less noisy than window fans. Whole-house fans are effective at cooling a house when the outside temperature is less than 80°F. Even if you live in a very hot climate, you could use a whole-house fan instead of your air conditioner during the early and late summer, and at night and in the early morning.

→*How does it work?* The large fan is usually installed in the highest ceiling that connects to your attic, pulling fresh air into your house through your (open) windows, sucking that air into the attic, and pushing it out through the roof vents. Obviously, you need to have adequate attic vents for a whole-house fan to work. A qualified electrician should install your whole-house fan. Don't use your air conditioner at the same time as your whole-house fan, and be sure to cover the fan tightly in the winter so warm air doesn't escape into your attic.

Low Cost

More Savings, Less Carbon Dioxide

Annual amount of *money* **saved** as a result of using a whole-house fan instead of an air conditioner for ⅓ of the summer, taking into account initial cost:

$28

Annual amount of CO_2 **not** *emitted* as a result of using a whole-house fan instead of an air conditioner for ⅓ of the summer:

825 lbs.

Assumptions: Average U.S. household uses 1,677 kilowatt-hours on central air-conditioning; whole-house fans use one-tenth of that energy; national average price of electricity is 8.16 cents per kilowatt-hour; whole-house fan costs $200 (including installation) and lasts 15 years.

Search For More Info

- www.easy2diy.com/tutorials/diy0119/index.asp—Learn in detail how to install a ridge vent in your attic. Go here, www.easy2diy.com/tutorials/diy0120/index.asp, to learn how to install a soffit vent.

- www.eren.doe.gov/buildings/home_fan.html—Check out this informative fact sheet about whole-house fans. Also, go here to learn all about solar attic fans: www.fan-attic.com/index.html.

- www.servicemagic.com—Go to this web site to get an estimate from a contractor of how much your ventilation project will cost or to find a prescreened contractor in your area. Select "Heating and Cooling" and then Fans to find a contractor near you who can install a whole-house or solar-powered fan.

Insulate Your Home

Each year, poor insulation in American homes causes the unnecessary emission of 67 million tons of carbon dioxide.

Overview. Some people shouldn't even bother turning on their furnaces—because their insulation is so poor that the hot air goes right through the roof! Heating and cooling your home represents about 50 percent of your annual energy bill, but up to 30 percent of that typically goes to heat (or air-conditioned air) that escapes through your ceilings, walls, floors, and ducts. Revamping your home's insulation is one of the best ways to improve the efficiency of your heating and cooling system. Besides saving money and energy, insulation will make your home feel more comfortable (in winter *and* summer) and block noise from the outside. With benefits like that, why wouldn't you want to add some insulation this weekend?

What You Should Know

- Heat will always flow to an area that's cooler. That's why, in the summer, hot outside air convects through your roof and walls to the cooler inside, and in the winter, warm air from your furnace pushes through the roof and walls to the cooler outside. Insulation, such as fiberglass, restricts heat flow, which means your furnace and air conditioner won't have to work as hard to keep your house comfortable.

- It's not true that insulating one part of your home, such as the attic, will just cause more heat to be lost through your windows. Insulation does not create any sort of pressure. The more you can bundle up your house with insulation, the better.

- If you're building a new home or renovating, however, be sure to put high-efficiency insulation into the walls and under the exterior siding, since these are the areas that are hardest to reach later.

- Insulation is rated by its ability to resist heat flow, called the R-value. The higher the R-value, the better the insulation. By putting new insulation on top of old insulation, you add the R-values of the two together. You'll find the R-value printed on the bag, the label, or the insulation itself.

- Your state or local government determines the minimum amount of insulation required for new homes or additions. When installing insulation into your home, make sure you achieve *at least* that amount. We recommend insulating your attic to at least R-38, your exterior walls to at least R-19, your crawl space and basement ceilings to at least R-11, and your heating ducts to at least R-6.

- *Climate Results:* If only one-fourth of Americans whose attics are insulated to levels below R-30 increased their insulation to R-38, they would prevent over 28 *million tons* of carbon dioxide from being emitted.

- *Money Matters:* Your home was insulated to the level that was required *at the time it was built.* But energy standards have gotten tougher as we've become more aware of our dwindling fossil fuel resources. If you have an older home and add insulation to meet today's standards, you could reduce your heating and cooling costs by 30 percent.

Easy Ways You Can Help

- *Insulate, insulate, insulate!* The more insulation your home has, the more hot air it will retain in the winter and air-conditioned air it will retain in the summer. Follow these five steps to improve your home's insulation.

Step 1: Determine your ideal amount of insulation.

The total amount of insulation your house should have depends on a variety of factors, such as climate and your heating fuel and system. To figure out how much insulation your house should have, go to www.simplyinsulate.com and select your state from the pull-down menu. To determine that ideal amount, however, you'll need to know how much insulation your home already has. An energy auditor can perform an insulation check, sometimes for free (see Tip 29), or you can check yourself.

Step 2: Check your current amount of insulation.

Either you or an auditor should check the following five areas of your home to see if they have insulation. If they do, you should try to determine its type and thickness. Check:

→ ***The attic,*** including its ceiling, walls, and floor.

→ ***Floors*** that lie above unheated areas, such as a basement or crawl space.

→ ***Ceilings*** of unheated basements and crawl spaces, as well as walls of any *un*ventilated crawl spaces.

→ ***Exterior walls*** (walls that separate a heated area from an unheated area, such as a wall between your heated kitchen and your unheated garage, or between your living room and the unheated outside). If your basement is *un*heated, you don't need to insulate its walls.

→ ***Heating Ducts.*** Most of the heating ducts that carry warm air from your furnace to the rest of your house run through unheated areas such as crawl spaces, basements, or attics before they reach the heating vents in your rooms. To prevent heat loss through these unheated areas in the winter (or heat gain from these uncooled areas in the summer), you should insulate the ducts themselves. If you have a boiler instead of a furnace, you should insulate any exposed hot-water pipes.

Step 3: Pick the type of insulation.

Use this table to help you decide what kind of insulation to buy:

Type of Insulation	Where Should You Install It?	What Is It Made Of?	How Is It Installed?	Benefits
Blanket	In the attic, or in any unfinished ceilings, walls, or floors	Fiberglass or rock wool, comes in batts or rolls	Fits between wall studs, floor joists, and beams	You can install it yourself.
Loose-fill	Inside existing walls or hard-to-reach places	Cellulose, polyurethane foam, fiberglass, or rock wool	Blown into small or large spaces with professional equipment	It insulates closed off and hard-to-reach areas.
Rigid	Basement and inside exterior walls	Extruded polystyrene foam or similar foams	Placed inside walls during new construction or renovation	It's highly insulative for such a thin board.
Reflective (a.k.a. multilayered radiant barrier)	Attic floor, attic walls, or directly under roof	Multilayered, foil-faced paper or cardboard	Fits between roof rafters, floor joists, or wall studs	You can install it yourself (see Tip 21 for more info).

Adapted from Oak Ridge National Laboratory's "Table 1. Types of Insulation."

Step 4: Decide who should do the job.

Only two people can insulate your home: a contractor or you. The ultimate decision will depend on your house, the type of insulation being added, and your skills. You might be able to handle the job in the attic, but only a professional can add insulation to your walls (by using special equipment to blow in loose-fill insulation). If you decide to go with a contractor, take the time to pick a capable one. To find a contractor that belongs to the Insulation Contractors Association of America, go to www.insulate.org and click on "Consumers."

Step 5: Install the insulation!

If you decide to go it alone, make sure you follow all the safety and fire prevention guidelines for the insulation you're adding. Go to the web sites listed in "Search for More Info" for detailed instructions.

→ ***The Attic.*** On your own, you can probably manage insulating your attic or any ceilings of unheated spaces using blanket or reflective insulation. In the attic, it is imperative that you plug all air leaks and take the necessary ventilation measures (see Tips 23 and 24) before you add insulation. Then, lay or staple the new insulation crossways on top of the old insulation for the best results, making sure you're not covering up any vents. Don't forget to insulate the attic door or hatch as much as or more than the rest of the attic.

→ ***The Ducts.*** You should insulate all accessible ducts using special duct-insulating wrap. If you can afford it, first seal the ducts' air leaks (see Tip 26). Never insulate the hot flue pipe of a gas appliance, and never insulate over a light fixture or other source of heat.

Low Cost
More Savings, Less Carbon Dioxide

Annual amount of *money **saved*** as a result of insulating a 600-square-foot attic to R-value 38 by yourself, taking into account the cost of the insulation:

$225

Annual amount of CO_2 ***not** emitted* as a result of insulating a 600-square-foot attic to R-value 38 by yourself:

2,149 lbs.

Assumptions: Based on average U.S. household usage of 57.7 million Btus and $561 per year on heating and cooling; typical attic is 600 square feet, insulating to R-38 reduces heating and cooling usage by 30 percent; blanket insulation costs 46 cents per square foot and lasts at least 15 years; 100,000 Btus = 1 therm; 1 therm emits 12 pounds carbon dioxide.

Search For More Info

- www.easy2diy.com/tutorials/diy0117/index.asp—This how-to guide shows you exactly how to insulate your attic. Go here, www.energyoutlet.com/res/ducts/insulating.html, to learn how to insulate your ducts.

- www.ornl.gov/roofs+walls/insulation/ins_01.html—This fact sheet tells you what to look for in a contractor (click on "If You Have It Done Professionally") and gives detailed installing instructions for insulation. Go to www.ServiceMagic.com (click on "Heating and Cooling" then "Insulation") to find a prescreened insulation contractor in your area.

- www.simplyinsulate.com—Check out this web site to find out which rebate or incentive programs are available in your state.

Leaky Ducts

Between 25 and 40 percent of the hot air your furnace pumps out escapes through leaky ducts before it can reach your home's rooms.

Overview. You could be paying hundreds of dollars a year to heat your crawl space. That's because of a very important yet often overlooked part of your heating and cooling system—the ducts. You probably don't think of them much because they're hidden behind your walls or in your crawl spaces. However, ducts are essential to your comfort. They carry the warm or cool air from your central furnace or air conditioner to the rooms all over your house, pumping it in through the vents in your floors or walls. Unfortunately, ducts are usually installed quickly and sloppily, causing them to leak air through poorly sealed seams or even gaps. As a result, much of the heated air in the winter or cooled air in the summer escapes into your crawl spaces or attic before it has a chance to reach your rooms. If your ducts were airtight, though, this air wouldn't escape, your furnace or air conditioner wouldn't have to run as long, and you would save a bundle of energy and money!

What You Should Know

- A typical U.S. home has 180 feet of ducts.

- If you seal your leaky ducts, your furnace or air conditioner will heat or cool your house more quickly and distribute the warm or cool air more evenly, and you'll greatly reduce the chances of dust, humidity, or gas fumes entering your heating or cooling system.

- A leaky duct system not only wastes energy but also can worsen the quality of your indoor air. For instance, a leaky duct system can cause harmful gases from gas appliances (such as a furnace, stove, or water heater) to be circulated to the rest of your house. A leaky duct system can also cause the air pressure in your home to decrease, sucking unwanted hot or cold air from the outside in.

- Periodically checking and resealing your heating ducts is always a good idea, but you should definitely do it when you're buying a new heating or cooling system. A new, superefficient furnace or air-conditioning system won't save you a penny if half the warm or cool air disappears through leaky ducts.

- *Climate Results:* Duct leakage in the United States is responsible for over 63 *million tons* of carbon dioxide being emitted each year—that's how much carbon dioxide we'd *prevent* from being emitted if we all resealed our ducts this year.

- *Money Matters:* Duct leakage costs Americans $10.2 billion each year in unnecessary heating and cooling costs. With that money, we could replace 537,000 gas-guzzling SUV's with superefficient, gas-electric hybrid cars.

Easy Ways You Can Help

- *Check your ducts for leaks.* Hop into that crawl space or attic, or bring a flashlight to your heating vents, and start looking for disconnected, kinked, or crushed ducts as well as any gaping holes. If there's a room in your house that never gets warm or cold enough, lift off the vents in that room and use a flashlight to look for a disconnected duct.

  To detect small leaks, turn your furnace or air conditioner on and run your hand along the duct to feel for escaping air. Also, look for dirt streaks near the ducts' seams or on the insulation covering the ducts. Dirt indicates an air leak where dusty air has been escaping for years. You may also find peeling duct tape, exposed metal, or ripped outer covers, all of which could indicate an air leak. Last, the most important place to check for leaks is where your ducts connect to your furnace.

- *Test your ducts.* Although disconnected ducts are easy to spot on your own, it's best to locate small duct leaks using professional testing equipment. Consider having a heating contractor or energy auditor conduct a duct blaster test, which pressurizes your duct system to tell you exactly which ducts are leaking and by how much. The total duct air leakage in older homes should be no higher than 12 percent, about half that for newer homes. Of course, the closer you can get your system to 0 percent air leakage, the better. Your contractor should also redo this test *after* he or she seals your ducts to make sure that the leaks have indeed been fixed.

- *Seal your leaky ducts.* You can get a contractor to seal your leaky ducts or you can try doing it on your own, as long as you know how to use the sealing materials safely (read all directions!) and are aware of your local building codes (which tell you what kind of sealant you can use, et cetera). First, rejoin any disconnected ducts using duct ties, draw-bands, or mechanical fasteners. Then seal any leaks or seams with water-soluble mastic caulking only. (See "Search for More Info.")

- *Don't use duct tape!* Believe it or not, regular old duct tape should *not* be used to seal ducts, since its adhesive will dry out over time, causing it to peel or fall off. Even UL-181 (approved by Underwriters Laboratories, Inc.) or foil-faced duct tape is just a temporary fix. Also, simply insulating your ducts will not fix air leaks since insulation stops only heat flow, not airflow.

- *Go high-tech.* There is a new, high-tech way to seal your ducts called Aeroseal. Here's how it works: A professional contractor closes all your vents and then blows adhesive particles, called Aeroseal, into your duct system. The particles stick to and fill up your leaks. It's as easy as that! This new invention is faster and more effective than sealing with mastic caulking or tape, and costs about the same as having a professional seal your ducts with mastic caulking. (See "Search for More Info.")

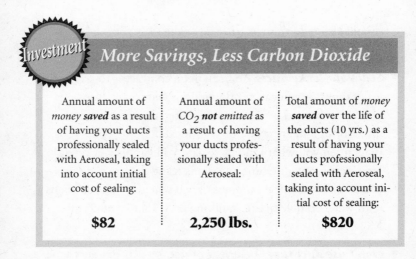

Investment

More Savings, Less Carbon Dioxide

Annual amount of *money* **saved** as a result of having your ducts professionally sealed with Aeroseal, taking into account initial cost of sealing:	Annual amount of CO_2 **not** *emitted* as a result of having your ducts professionally sealed with Aeroseal:	Total amount of *money* **saved** over the life of the ducts (10 yrs.) as a result of having your ducts professionally sealed with Aeroseal, taking into account initial cost of sealing:
$82	**2,250 lbs.**	**$820**

Assumptions: Based on average U.S. household that consumes 57.7 million Btus per year and spends $561 per year on heating and cooling; duct sealing reduces total heating and cooling costs by 32.5 percent; Aeroseal, including a duct blaster test before and after sealing, costs $1,000 and lasts at least 10 years; 100,000 Btus = 1 therm; 1 therm emits 12 pounds carbon dioxide.

Search For More Info

- www.eren.doe.gov/buildings/documents/pdfs/27630.pdf—Check out this in-depth report from the Department of Energy on ducts and energy efficiency.

- www.homeenergy.org/consumerinfo/ducts/index.html—This great do-it-yourself guide has photos that show you exactly how to seal your ducts.

- www.aeroseal.com—Go here for more information about Aeroseal duct sealing and to find an Aeroseal contractor near you.

Replace Your Windows

Energy Star–labeled superwindows can cut your total annual energy bill by 15 percent, making them one of the single best ways to reduce energy usage.

Overview. Windows have always been the Achilles' heel of heating and cooling a house. That's because in winter precious heat from your furnace passes through the glass to the outside, and in summer heat from the outside presses into your air-conditioned home. At the same time, no one wants to give up having sunlight pour into the house or being able to see the outdoors. Well, lucky for you, a solution has arrived. Thanks to modern technology, we no longer have to sacrifice our energy bills in order to have a good view. New superwindows with low-emissivity coatings keep summer heat out and winter heat in. If new windows are not within your budget this year, don't worry. There are many other ways you can make your windows more efficient.

What You Should Know

- The best superwindows insulate *five* times better than single-paned windows. Plus, today's Energy Star windows save 50 percent more energy than a typical 10-year-old window.

- If you're planning on buying new windows anyway, the most energy-efficient models will cost you only slightly more, but they'll provide significant energy savings for decades to come.

- Single-pane, unglazed windows are the most inefficient windows out there. In severe weather, a single-pane window loses heat 10 times faster than your home's walls do.

- The biggest breakthrough in recent window technology is the low-emissivity or low-E coating. This invisible coating of inert gas molecules restricts the transfer of heat, blocking up to 70 percent of the sun's heat that would normally pass through glass, yet allowing 100 percent of the sun's light to pass through. The result? An insulated window that's guaranteed to last for 50 years.

- The efficiency of a window is determined by four things:

  1. The U-value tells you how much heat can pass through the glass *and the frame* (ranges from 1.10 to 0.33, the lower the better). Make sure you know the U-value for the entire window and not just for the glass.
  2. The solar heat gain coefficient (SHGC) is the amount of solar energy that can pass through your window and then turn into heat (ranges from 0.4 to 0.9; hot climates want low SHGC's, cold climates want high ones).
  3. The visible transmittance (VT) is the percentage of visible light that can pass through a window (ranges from 0.0 to 0.9; a higher VT means more light gets through).

4. The air leakage rating measures how much air leaks around a window at a certain wind speed (the lower the air leakage rating, the better; look for a rating of 0.01 to 0.06 cubic feet per minute per square foot).

- *Climate Results:* If just one family installed storm windows this winter, they would prevent over 1,000 pounds of carbon dioxide from being emitted.

- *Money Matters:* About 33 percent of the heat your furnace produces in the winter is lost as the heat convects through windows and doors to the colder outside. For the average household, that adds up to about $140 a year.

Easy Ways You Can Help

- *First things first.* A brand-new superwindow isn't worth a dime if air is still leaking in (or out) around its edges. See Tip 23 to learn how to make your windows and doors airtight.

- *Make a sunroom.* Using passive solar energy, which means designing your home to take advantage of maximum sunlight, can reduce your combined heating and lighting bill by up to 40 percent. If you're building a new home or remodeling an old one, be sure to add a lot of big, south-facing windows and fewer north-facing ones (where the sun will hardly ever shine in). In North America, the sun is lower in the winter and more overhead in the summer, so you'll want to place large windows lower to gain heat as well as light in the winter. Place smaller windows, which can be more easily shaded, higher up so they don't overheat your house in the summer.

- *Get superwindows!* If your windows are old or single-pane, seriously consider investing in the newest, triple-pane super-windows, which have inert argon gas between their three panes to add extra insulation, as well as two low-emissivity coatings. Be sure to look for the Energy Star label, which means the windows are extraefficient.

    First, go to the map on Energy Star's web site, www.energystar.gov/products/windows/#climate, to see what kind of superwindow is right for your climate. In general, if you live in a hot climate, you'll want to block solar heat gain; if you live in a cold climate, you'll want to allow for maximum solar heat gain; and if you live in a mixed climate, you'll want to block heat gain in the summer and allow for it in the winter. Then go here, www.energystar.gov/stores/storelocator.asp, to find a store that sells Energy Star windows near you.

- *Stop summer heat gain.* Does too much sunlight streaming through your windows cause your house to overheat in the summer? If so, think about buying windows with a low-E coating that is also spectrally selective, meaning that it lets visible light pass through but reflects the long-wave infrared rays that turn into heat. Spectrally selective low-E coatings also prevent 75 percent of ultraviolet rays from passing through your windows and damaging your skin, furniture, and artwork. Another alternative would be to buy tinted windows, although they aren't as good at reducing heat gain and also make your rooms darker. Last, you can install temporary plastic window films (see "Put on a Plastic Film") or sun-control screens (see Tip 22) to reduce summer heat gain.

- *Install storm windows.* If brand-new windows are not within your budget, see if storm windows are. These extra panes made of glass or clear, strong plastic fasten with clips or screws on either the inside or outside of your permanent window frame. They are typically used only in the winter. Storm windows reduce air leaks, water condensation, and frost buildup and increase a window's insulating ability. Some new storm windows even have low-emissivity coatings. Storm windows are particularly effective on single-pane windows.

- *Put on a plastic film.* If storm windows aren't in your budget, your best bet is to pick up a kit of inexpensive clear plastic sheets from your hardware store. The latest ones have low-emissivity coatings and can reduce winter heat loss through windows by up to 40 percent! You can easily apply them to the insides of your windows for the winter, take them off for the summer, and reuse them. You can also apply different films in the summer to help *reduce* heat gain. Window films usually last 8 to 10 years, but they're easily damaged by weather and may reduce your visibility somewhat.

- *Look at the stars . . .* through a skylight! Skylights (windows in your roof) can eliminate the need for electric lighting during the day, let in heat during the winter, and provide ventilation in the summer (if they open). Be careful, though—if left uninsulated or uncovered, skylights can lose tremendous amounts of energy. Most skylights today come with remote-control blinds or panels that you can lower to keep in heat at night or prevent heat gain in the summer.

- **Look to the future.** Superwindows are amazing, but "smart" windows are going to blow you away. Imagine flicking a switch and having your windows immediately become tinted, then flicking it again and having them go back to clear. These technologies are still being tested, but look for them soon. Do your homework, though—some are energy savers while others are energy wasters.

More Savings, Less Carbon Dioxide

Annual amount of money **saved** as a result of replacing your windows with Energy Star superwindows, taking into account the initial cost of new windows:	Annual amount of CO_2 **not** emitted as a result of replacing your windows with Energy Star superwindows:	Total amount of *money* **saved** over the life of the windows (50 yrs.) as a result of replacing your windows with Energy Star superwindows, taking into account the initial cost of new windows:
$68	**1,864 lbs.**	**$3,400**

Assumptions: Based on average U.S. household total annual energy costs of $1,386 and consumption of 103.6 million Btus, where superwindows reduce overall energy usage by 15 percent and last 50 years; seven three-by-four-foot windows cost $7,000 (including installation); 100,000 Btu = 1 therm; 1 therm emits 12 pounds carbon dioxide.

Search For More Info

- www.energystar.gov/products/windows—Check out Energy Star's web site for superwindows.

- www.dulley.com/gtopics.shtml—Click on "Windows Improvements," then scroll down to read Bulletins 460 and 617, both about plastic window films, and 563 about storm windows.

- www.consumerenergycenter.org/homeandwork/homes/inside/windows/future.html—Check out this web site to learn all about the smart windows that are currently being tested.

- *Residential Windows: A Guide to New Technologies & Energy Performance*, second edition, by John Carmody. Check out this in-depth book about the latest in window technologies.

Invest in a New Furnace or Air Conditioner

Today's top air conditioners are 50 percent more efficient than ones made 20 years ago. New furnaces are 45 percent more efficient.

Overview. Each year approximately 45 percent of your annual energy bill goes toward heating and cooling your home. That's why it's so important to buy an energy-efficient system (and then maintain it so it will continue to operate efficiently). If your furnace or air conditioner is getting old or worn out, you should investigate the possibility of upgrading. It could be a very profitable and energy-friendly option.

What You Should Know

- A furnace lasts about 20 years, a central air conditioner lasts about 15 years, and a room air conditioner about 10 years. When we're talking *that* long, isn't it worth it to make a wise investment?

- Approximately one-third of all home air conditioners in the United States are oversized by *50 percent*. In other words, they're twice as big as they need to be for the house in which they're used. Running a smaller air conditioner for a longer period of time is more energy-efficient—and more effective at cooling your house—than running an air conditioner that's too large for the space.

- A typical household spends about $420 a year on heating and at least $140 on air-conditioning.

- Improper installation is responsible for up to 50 percent of a furnace's heat loss, which is why it pays to hire a qualified, experienced contractor.

- The U.S. government has created two rating systems to help you identify the most efficient furnaces and air conditioners on the market. For furnaces, look for the annualized fuel utilization efficiency (AFUE) percentage, which is the ratio of the amount of fuel that enters your furnace to the amount that is actually converted into useful heat. The higher the AFUE, the better (a 100 percent AFUE rating would be perfectly efficient). The AFUE percentage of a furnace bought 20 years ago is about 50 percent. Today's minimum AFUE requirement for new furnaces is 80 percent, but the best furnaces today have 96 percent AFUE's!

- For central air conditioners, look for the seasonal energy efficiency ratio (SEER), and for room air conditioners, look for the energy efficiency ratio (EER). These ratios reflect the amount of heat removed from the air per watt of electricity used. The higher the EER or SEER, the better. A 20-year-old central air conditioner will have a SEER rating of about 5.0, ones that were built after 1990 have SEER's of 8.0 or higher, while today's top models have SEER's of 16.0! You will find both AFUE and EER or SEER ratings on EnergyGuide labels (see Tip 14), which are required on all air conditioners, furnaces, and boilers.

- *Climate Results:* Today, more than one-fourth of home furnaces in the United States are over 20 years old. If every household that had one of these clunkers upgraded to an Energy Star furnace, they would collectively prevent 273 *million tons* of carbon dioxide from being emitted each year.

- *Money Matters:* Upgrading from a central air conditioner that is 50 percent oversized to a properly sized central air conditioner can save you $85 or more a year.

Easy Ways You Can Help

Make the Decision . . .

→*Is it time?* If your furnace or boiler is 15 years old or older, worn out, or too big for your house, right now may be a good time to upgrade to a superefficient model. Upgrading is an especially good idea if your gas furnace has a pilot light instead of electric ignition. Central air conditioners will last about 15 years, but if yours was poorly installed or poorly maintained (see Tip 20), it might make sense to upgrade now.

→*Get what's best for you.* Climate is the major factor that determines the type and size of furnace or air conditioner you choose. Central air-conditioning systems are best suited for hot climates, while room air conditioners are better for more moderate climates. A superefficient furnace (96 percent AFUE) makes sense only if you live in a cold climate, whereas a moderately efficient furnace (80 to 90 percent AFUE) or heat pump is best in milder climates.

→ ***Mix it up.*** Using more than one type of heating system is becoming more and more popular as a way to cut heating costs. For example, environmentally friendly heating options such as solar panels, advanced woodstoves, or advanced gas fireplaces can now be used as the primary heating source, leaving your central heating system as backup. Or, if you're building an addition, installing a separate heating system, heat pump, or room air conditioner for that area may be cheaper than trying to integrate it into your existing system.

Go Shopping . . .

→ ***Size it right.*** When air conditioners are too large for their cooling space, they can't remove humidity effectively, causing a clammy, damp environment. They also turn on and off too often, causing the compressor to wear down faster. An oversized furnace is similarly inefficient. Also, don't forget that larger units cost more initially and cost more to operate. By contrast, a unit that's too small will not be able to cool or heat your home effectively and will waste energy by running continuously. It is imperative that your contractor size your air conditioner and furnace exactly right—which brings us to . . .

→**Finding Mr. or Ms. Right.** A good contractor should size your new furnace or air conditioner on the basis of a number of factors, including the size of your house, the number of windows you have, the amount of insulation in your walls, the color of your roof, and so on. To determine the final sizing, he or she should use a complex formula created by a reputable organization such as the Air Conditioning Contractors of America (ACCA).

→**Go-go gadgets.** Choose both furnaces and air conditioners that have programmable thermostats and lights that go on when the filters need to be replaced. Also, seriously consider investing in a zone heating and cooling system, in which a specific thermostat controls the heating or cooling of each room or area.

→**Look for the Energy Star logo.** The Energy Star logo applied to central and room air conditioners and furnaces guarantees superhigh efficiency thanks to better compressors, fan motors, and heat exchangers. Energy Star air conditioners and furnaces usually cost more initially, but they will quickly pay for themselves (and then some!) thanks to lower operating costs. Also, a superefficient furnace or air conditioner will increase the resale value of your home. Buy the most efficient Energy Star furnace or air conditioner you can afford, which might be easier thanks to the various rebate and grant programs offered by utilities and state governments.

Bring It Home . . .

→ ***Install correctly.*** Unfortunately, if your top-of-the-line, 96
percent efficient Energy Star furnace is not installed properly, it could end up being just as inefficient as the 20-year-
old one you're replacing. Before you install your new
furnace or central air conditioner, you should seriously consider resealing your ducts so they're airtight (see Tip 26).
Also, if you're planning on adding insulation (Tip 25) or
new windows (Tip 27) anytime soon, do that first, since it
will allow you to buy a smaller (and therefore less expensive) furnace or air conditioner.

Investment **More Savings, Less Carbon Dioxide**

Annual amount of *money **saved*** and CO_2 ***not*** emitted, as well as the total
amount of money saved over the life of the product, as a result of upgrading
from a 15-year-old furnace and 10-year-old central air conditioner to Energy
Star models (taking into account the initial cost of the new models) . . .

Energy Star Furnace:	*Energy Star Air Conditioner:*
$57/Yr.	**$31**/Yr.
11,916 lbs. CO_2/yr.	**3,969** lbs. CO_2/yr.
$1,424/over 25 yrs.	**$465**/over 15 yrs.

*Assumptions: Energy Star furnace has 96.6 AFUE, costs $2,500 (including installation), lasts 25 years,
saves 993 therms of natural gas per year; 100,000 Btus = 1 therm; 1 therm emits 12 pounds carbon
dioxide; old furnace has 60 AFUE; average U.S. household spends $425 per year on heating; Energy
Star central air conditioner has 16 SEER, costs $2,500 (including installation), lasts 15 years, saves
2,420 kilowatt-hours per year; old central air conditioner has 8 SEER.*

Search For More Info

- http://yosemite1.epa.gov/estar/consumers.nsf/content/ hvac.htm—Go here to learn all about Energy Star furnaces and air conditioners and to find out where you can buy them.

- www.pge.com/003_save_energy/003a_res/heating/heating. shtml—Go here to learn more about correctly sizing your new furnace or air conditioner and selecting a good contractor.

- www.servicemagic.com—Go here to find a prescreened heating and cooling contractor in your area. Select "Heating and Cooling" from the list of categories.

The Advantages of an Audit

The steps suggested by an energy auditor, such as adding insulation, sealing air leaks, or buying a new fridge, usually pay for themselves in energy savings within three years or less.

Overview. This book is for the average person, but of course, you're not average. Everyone owns a different home, lives in a different climate, and enjoys a different lifestyle. That's why, in order to get the most out of these 51 tips, you're going to need some personal attention. And the best way to get that is by having an energy audit. An energy audit is simply an assessment of how your household currently uses energy, and how that usage can be improved to save the most energy and the most money possible. An auditor will identify your home's weak spots and show you how to fix them, whether it's the insulation, the windows, the air leaks, or the 20-year-old refrigerator. We know you're not average—and an energy audit will tell you exactly how you're unique.

What You Should Know

- If home and business owners were to implement energy-efficiency measures today (such as those suggested by an energy auditor), in the next 20 years we would have saved 25 percent more oil than we could extract from the Arctic National Wildlife Refuge in the next 60 years.

- Getting an energy audit will tell you not only how to save significant amounts of money and energy but also how to increase the comfort and life span of your home by reducing noise, moisture, squeaks, creaks, and leaks.

- Receiving a high energy-efficiency rating on your house from a licensed auditor will increase its resale value because it means that your house is less expensive to operate than other homes. If you're buying a new home, be sure to ask what its energy-efficiency rating is.

- If you're remodeling a current home or building a new one, remember that it's *always* cheaper to include energy-efficient features from the get-go rather than add them in a few years down the road. Furthermore, the extra initial cost of these features will usually be repaid in energy savings within just a few years.

- *Climate Results and Money Matters:* If just 100,000 of American households received energy audits this year and followed all their auditors' recommendations, they would collectively save around $64 million in energy costs and prevent

1.2 *billion pounds* of carbon dioxide from being emitted every year.

Easy Ways You Can Help

- ● *Take an on-line energy audit.* Many utility companies, government agencies, and independent web sites now offer helpful and accurate on-line energy audits for free. With some, you can enter figures from your previous energy bills to get an extremely accurate audit. With others, you can manipulate your choices to compare the monetary and energy savings of different behavioral changes and invest-ments. These web sites are generally quick and easy to use, offering personalized energy-reducing tips. Many also have links to in-depth fact sheets on related topics, toll-free num-bers you can call if you have questions, links to on-line stores that sell energy-efficient appliances and products, and lists of energy-conscious contractors in your area.

- ● *Here are a few on-line energy audits we recommend,* although new and improved ones pop up all the time. If your utility offers one that can make calculations using your past bills, try that first.

  1. http://homeenergysaver.lbl.gov: After entering your zip code, you can customize this on-line audit for your own home.
  2. www.ase.org/checkup/home/main.html: This quicker but more generalized audit focuses primarily on how much

you can save on your energy bills. Select different efficiency levels to see how certain actions will affect your energy bill.

3. www.energyguide.com/audit/HAintro.asp: Punch in your zip code, click on "In-Depth Analysis," then click "Begin." This web site makes you register with an e-mail address, but it's generally helpful and easy to use.

- *Take a stroll through your house.* You can do your own energy audit simply by inspecting your home. Look for obvious air leaks (see Tips 23 and 26), check to see how much insulation you have (see Tip 25), and examine your appliances: How old are they? Are they functioning well? (see Tip 14).

- *Get a professional energy audit.* In a professional audit, a qualified, unbiased technician comes to your house and uses specialized equipment to determine how energy-efficient your house is. He or she then recommends steps you can take to reduce your energy usage (and energy bills) while maintaining or even improving your level of comfort. A good auditor will first inspect the outside of your home, then do a room-by-room inspection, see how much insulation you have, examine your utility bills, ask about your behavior (Is anyone home during the day? How many people live in the house? and so on), and carry out technical tests.

What Kind of Technical Testing Will Be Done?

An energy auditor might use specialized equipment, such as blower doors, furnace-efficiency testers, surface thermometers, thermographic scans, or computer simulation software to test the efficiency of your home. These tests detect inefficiencies that you wouldn't be able to see during a simple walk-through. (See "Search for More Info.")

How Can I Find an Auditor?

Call your local utility company to see if they offer energy audits for free or for a subsidized fee. If your utility doesn't offer audits, contact your state Energy Office (see "Search for More Info") for a list of auditors in your area, or check the yellow pages under "Energy" or "Energy Management." An audit can cost anywhere from $50 to $500, so choose your auditor carefully. Ask your neighbors for recommendations, look for a local contractor, make sure he or she is licensed and insured, ask for quotes (in writing) from at least two contractors, and request references.

Investment

More Savings, Less Carbon Dioxide

Annual amount of *money* **saved** as a result of following an energy auditor's suggestions (see assumptions), taking into account the initial costs of the audit and the energy efficiency measures:

$203

Annual amount of CO_2 *not emitted* as a result of following an energy auditor's suggestions (see assumptions):

12,595 lbs.

Assumptions: Having the audit, sealing your ducts, and buying compact fluorescent lightbulbs, a programmable thermostat, a water heater wrap, and a small Energy Star refrigerator cost a total of $839; save $287 per year; all measures last (at least) 10 years; U.S. average amount of carbon dioxide prevented by an energy audit = 12,595 lbs.

Search For More Info

- www.servicemagic.com—Use this search engine to find a prescreened energy auditor (or any type of contractor) near you. Search in the category "Appraisers and Inspectors," under "Moving and Real Estate."

- www.eren.doe.gov/consumerinfo/refbriefs/ea2.html—Check out this great fact sheet on do-it-yourself audits and what to expect from professional audit testing.

- www.naseo.org/members/states.htm—Click on your state to find the address, phone number, and Internet address of your state's Energy Office.

Your Backyard

Save with Shade

Adding energy-efficient landscaping to your home can reduce your air-conditioning costs by up to 50 percent.

Overview. When was the last time someone told you to go play in the dirt? Well, that's exactly what we want you to do. By planting the right kinds of plants in the right places around your home, you can *significantly* reduce your air-conditioning bill in the summer and your heating bill in the winter. But the perks don't stop there! With energy-efficient landscaping, your house will be more comfortable to live in, you won't have to water or care for your plants as much, you'll create a habitat for birds and wildlife, increase your property value, protect your home from the elements, reduce noise and air pollution, and best of all, live in a more beautiful home. With a list as long as that, what are you waiting for?

What You Should Know

- Shading your outdoor or window air-conditioning unit can improve its efficiency by 10 percent.

- Vines growing on a trellis can reduce the surface temperature of the wall the trellis is attached to by 40°F. Talk about lowering your air-conditioning costs!

- *Climate Results:* If 100,000 people planted shrubs or put up a wooden screen to shade their air conditioners, they would prevent 14,000 *tons* of carbon dioxide from being emitted every year.

- *Money Matters:* A typical energy-efficient landscaping project will pay for itself in energy savings within eight years.

Easy Ways You Can Help

- *See where it's sunny.* The first step in energy-efficient landscaping is determining where and how much sun hits your house at different times of the day. Then, depending on your climate, you should take steps to block the sun, let it shine through, or a bit of both.

- *Know your climate.* There are four major climates in the United States. Here's what your landscaping priorities should be depending on your climate:

  →*Hot-arid:* Use high-canopied trees to create maximum shade on your roof, walls, and windows. If you don't use air-conditioning, allow breezes to reach your house. If you do use air conditioning, deflect breezes away from your house.

  →*Temperate:* Use deciduous trees (ones that lose their leaves in winter) to shade as much sun as possible during the summer and maximize solar heat gain during the winter. Put up a windbreak to deflect winter winds yet allow summer breezes to reach your home.

→**Hot-Humid:** Allow summer breezes to reach your house. Create as much shade as possible in the summer with deciduous trees, which will also allow the sun to reach and warm your home in winter.

→**Cool:** Plant a thick windbreak to deflect winter winds. Make sure sunlight can reach south-facing windows in the winter. Plant deciduous trees if needed to prevent overheating in the summer.

- **Grow vines.** A trellis covered with vines not only is beautiful but also can shade (and therefore cool) a wall or patio very effectively. With a deciduous vine, the sun will be able to shine through to warm your home in winter. Because vines are fast-growing, they will begin providing shade within one year. Ask your local nursery which vines are right for your climate.

- **Plant bushes.** Small- to-medium-size bushes and trees can effectively shade your walls, especially when planted on the west side of your house, where low afternoon sun shines in. When planted along a sidewalk, driveway, or parking area, they block the heat that bounces off those flat surfaces and onto your home. When planted one foot from the edge of your house, small trees or bushes provide cooling shade in the summer and extra insulation in the winter. Also, larger trees are vital providers of shade for your house (see Tip 31).

- **Go lower.** Don't forget the ground cover! Ground cover planted around the edges of your house or along your driveway and walkways will cool these surfaces in the summer (by reducing heat radiation from the ground) and insulate them in the winter. Three of the most popular ground covers are ivy, pachysandra, and blue rug juniper.

- **Shade your air conditioner.** Plant small shrubs or build a wooden screen to shade your air conditioner, but leave about two feet of clearance on the sides and four feet above the unit so air can still flow easily. If possible, install the unit on the northern (the shadiest) side of your house. Last, try to minimize the amount of debris, such as dirt, leaves, or twigs, on or around your air conditioner. A dusty and dirty air conditioner doesn't work as efficiently as a clean one.

Low Cost

More Savings, Less Carbon Dioxide

Annual amount of *money* **saved** as a result of covering your east and west walls with trellises of vines, taking into account the initial cost of the trellis and vines:	Annual amount of CO_2 **not** *emitted* as a result of covering your east and west walls with trellises of vines:
$13	**548 lbs.**

Assumptions: Based on an average U.S. household that uses 5.7 million Btus and spends $140 per year on air-conditioning; vines on east and west walls reduce cooling costs by 20 percent; trellis and vines cost $190, last for 13 years, and cover two 400-square-foot walls (east and west).

Search For More Info

- www.xeriscape.org/xeriscape.html—Go here for more information on energy-efficient landscaping.

- www.dulley.com/gtopics.shtml—Click on the "Landscape/ Gardening" link, then scroll down to Bulletin 701 to read why shrubs and bushes are energy-efficient, and Bulletin 765 to learn more about vines and energy efficiency.

- www.cometobuy.com/directgardening/ProdMenu.asp—At this web site you can buy flowers, bulbs, trees, vines, and ground cover directly on-line.

Trees: Nature's Air Conditioners

An average-size tree soaks up 50 pounds of carbon dioxide through its leaves each year. That means it will store approximately one ton of carbon dioxide during its lifetime.

Overview. Everyone knows that trees provide shade. But did you know that they actually cool the air too? During photosynthesis, the large amount of water that was sucked up through your tree's roots is released into the air through its leaves—this is called transpiration. As this water evaporates, it cools the air surrounding the tree, similar to the cooling effect we humans experience when we sweat. In the summertime, an average-size maple tree can release more than 50 gallons of water in one hour. If that maple were close to your home, your air-conditioning costs would be greatly reduced. In fact, one maple tree has the cooling output of one large window air conditioner! That's why we call trees nature's own air conditioners.

There's another cool thing about photosynthesis that we should mention. During photosynthesis, plants' leaves absorb carbon dioxide from the air, and with the energy they get from sunlight, they are able to break it down into carbon and oxygen. Plants don't need the oxygen, so they release it, giving us fresher air to breathe. They do, however, need the carbon, which they store for the rest of their lives. In other words, trees and plants not only soak up and store the most prevalent global warming gas in our atmosphere—carbon dioxide—but they also provide us with fresh oxygen.

What You Should Know

- Although trees soak up carbon during their lifetime, when they die that carbon is released and turns back into carbon dioxide. In other words, planting trees to prevent global warming just *postpones* the problem. Planting more trees does *not* mean we can keep producing the same amount of carbon dioxide we've been emitting. In order to make up for the damage that's already been done, we need to plant trees *and* reduce our carbon dioxide emissions.

- A deciduous tree is one that loses its leaves in the winter. These trees are ideal for energy-efficient landscaping because they provide shade in the summer but, by losing their leaves in fall, allow the sun to warm your house in the winter. Some examples are green ash, red oak, sugar maple, and white elm.

- Think you have to transplant a 30-foot maple in order to have an impact on your cooling bills? Think again. A 6- to 8-foot deciduous sapling will begin shading your house within a year.

- Because of transpiration and shading, the air surrounding a tree is, on average, 9°F cooler. The Department of Energy has found that tree-shaded neighborhoods are up to 6°F cooler than neighborhoods without trees.

- Over their lifetime, one acre of trees soaks up as much carbon dioxide as a car would emit if someone were to drive it all the way around the world. Meanwhile, more than 35 *million acres* of rainforest are permanently destroyed every year.

- Most people think trees should shade windows, but it's just as important to shade your walls from the sun's forceful heat. This is especially true if your house is made of brick or stone.

- *Climate Results:* Putting up a windbreak of evergreens on the north side of a house reduces winter heating bills by an average of 33 percent. If all the households in the northern plains states (North Dakota, South Dakota, Nebraska, Wyoming, and Montana) planted windbreaks of evergreen trees, they could prevent nearly 2 *million tons* of carbon dioxide from being emitted every year.

- *Money Matters:* According to the Department of Energy, planting three trees strategically around your home will reduce your annual heating and cooling costs by an average of 40 percent.

Easy Ways You Can Help

- *Make some shade.* During the summer, deciduous trees can provide excellent shade for your walls, windows, roof, driveway, sidewalk, and patio while allowing the sun to warm your house in the winter after the leaves fall off. Plant these trees on the south and west sides of your house to block the most summer sun and provide the most winter warmth. For year-round shade, plant coniferous (evergreen) trees. See if your local utility company offers rebates for planting shade trees (shade trees benefit utilities because they help reduce peak-hour air-conditioning demands).

- **Block the wind.** Plant tall evergreen trees in a row (like a fence) on the north and northwest sides of your house to block strong winter winds, sending them up and over your house. Plant a row of shorter evergreen shrubs or bushes in front of the windbreak to stop wind and snow from blowing through the tree trunks toward your house.

  When buying either a deciduous or an evergreen tree, always take into account the tree's full-grown height, diameter, shape, and branch spread as well as its growth rate, winter durability, and life span. Also, be sure to ask your nursery about which trees are native to your climate and which ones will best serve your purpose (shading, wind blocking, et cetera).

- **Plant trees in your neighborhood.** Washington, D.C.'s heavy tree coverage has decreased by 64 percent over the last twenty years, and New York City's urban forest shrank by 175,000 trees during the 1990s. Meanwhile, thanks to a strict tree ordinance, residents of Gainesville, Florida, are enjoying air-conditioning bills that are on average $125 less than their friends' in the nearby town of Ocala, which doesn't have a strict tree ordinance. No matter where you live, ask your local government representative today about planting more trees in your neighborhood and fighting for stricter tree ordinances.

- *Plant trees in the rainforest—for free!* Just by clicking on one of the web sites listed here, you can save a part of the rainforest. Here's how it works: Companies buy the rainforest land in return for having you see their advertisements. You don't have to buy or sign up for anything. The worst that will happen is that an ad (which you can close immediately) will pop up on your screen. Make one of these web sites your browser's default home page so that donating rainforest land is the first thing you do every morning. If every household in the United States donated one tree to be planted in the rainforest this year, those trees would soak up 105 *million tons* of carbon dioxide over their lifetimes!

  *Note: For all three of these web sites, you can click only once per day.

  1. www.ecologyfund.com—Click once and save 105 square feet of rainforest from being clear-cut.
  2. www.therainforestsite.com—Click once and save 11.4 square feet of rainforest.
  3. www.clicktheplanet.org—This web site lets you save the most land with the fewest clicks. The catch: It's in Italian. Don't worry, though, you'll be able to follow it. Click on the lion cub to donate two square meters of African savanna, or click on the macaw to donate 20 square meters of Amazon rainforest. At the page you will see after you donate, hold your mouse over the ads. After a few seconds, when a pop-up message appears, click the ad to save even more land.

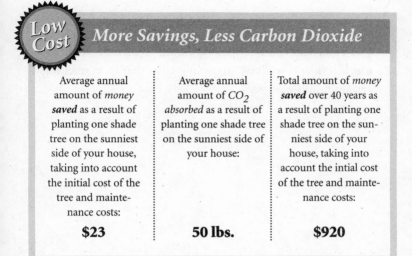

Low Cost

More Savings, Less Carbon Dioxide

Average annual amount of *money* **saved** as a result of planting one shade tree on the sunniest side of your house, taking into account the initial cost of the tree and maintenance costs:	Average annual amount of CO_2 *absorbed* as a result of planting one shade tree on the sunniest side of your house:	Total amount of *money* **saved** over 40 years as a result of planting one shade tree on the sunniest side of your house, taking into account the intial cost of the tree and maintenance costs:
$23	**50 lbs.**	**$920**

Assumptions: Based on average U.S. household that spends $421 per pear on heating, $140 per year on cooling; one tree absorbs one ton of carbon dioxide over 40 years; one tree reduces annual heating costs by 4 percent, cooling by 12 percent; installing a 13-foot sugar maple costs $127, maintenance (watering, fertilizer every other year) averages $8 a year.

Search For More Info

- www.powerhousetv.com/savingenergy/diy/tree.php3—This is a great do-it-yourself guide on choosing and planting the right tree for your home.

- www.conservation.state.mo.us/conmag/1998/03/5.html— Check out this web site to learn exactly how to plant a windbreak of evergreen trees.

- www.americanforests.org—American Forests can help you start a tree-planting campaign in your town. Use their CITYgreen software to make accurate predictions about how much residents will save in cooling costs thanks to planting a certain number of trees.

Green Plants, Less Water

Grass requires more water and maintenance than any other plant in your yard.

Overview. Water is a precious resource, but unlike fossil fuels, it isn't going to run out. It may become polluted and undrinkable, and there may not be enough freshwater in the world to sustain our runaway population, but as long as the Earth is around, water will be too. So why are we talking about water in a book about global warming? Because it takes energy (usually from the burning of fossil fuels) to collect, purify, and then transport the water that comes so effortlessly out of your faucet. Even more energy is used if that water has to be heated. Finally, energy is needed to recollect, repurify, and retransport the wastewater that so conveniently goes down your drain. Until the water we use for drinking and landscaping is transported and purified using renewable energy sources, such as solar or wind power, every drop that we use or drink is contributing to global warming.

What You Should Know

- Gardening is America's most popular hobby.

- Less than *3 percent* of the Earth's water is freshwater that we can drink. Of that 3 percent, only three-one-thousandths is available to us—the rest is locked up in glaciers, ice caps, and the Earth's core.

- Placing mulch around your plants, bushes, and trees can reduce your watering needs by 40 percent.

- Think homes don't use a lot of water compared with businesses or, say, golf courses? Think again. Forty-seven percent of *all* the water in the United States goes to residential homes. And half of the water a household uses each year goes to outdoor activities, such as watering your lawn or landscaping.

- Grass clippings are 75 to 85 percent water, which is why it's a good idea to leave them on your lawn after you mow (see the following section).

- A finely ground mulch retains more moisture in the soil than a mulch made up of larger pieces.

- For each hour of use, a gasoline-powered lawn mower is 11 times more polluting than a car. All the gasoline-powered lawn mowers in the United States emit more than two *million tons* of carbon dioxide a year.

- ***Climate Results:*** A lawn and landscape designed to conserve water requires 50 percent less water than a regular lawn. If the 6 million people who buy new homes every year designed water-conserving landscaping, they would collectively save 250 *billion gallons* of water and prevent 2.2 *million tons* of carbon dioxide from being emitted—each year.

- *Money Matters:* Planting native plants and grasses instead of nonnative ones can reduce your watering and maintenance costs by up to 85 percent.

Easy Ways You Can Help

- *Don't choose needy plants.* Make sure that every tree, bush, grass, and flower you plant in your yard is native to your climate. It will survive better and need much less care and water. And less water pumped through your hose means more electricity saved (which means fewer carbon dioxide emissions at the power plant!). If you live in a warm climate, use warm-weather grasses like Bermuda or zoysia, which require up to 75 percent less water than cool-weather grasses. Go here, http://gardening.about.com/cs/natives/index.htm— and click on your region to see a number of web sites that list plants and flowers native to your area. Also, ask your local nursery about which grasses are native to your area.

- *Think ahead.* Different types of plants require different amounts of water to survive, so try to plant things in groups according to how much water they need. That way you can water certain areas for specific amounts of time. Also, put plants that need a lot of water at the bottoms of slopes. Use as little concrete as possible to reduce runoff—try using porous concrete instead. Last, direct your gutters toward areas of your lawn that need water.

- *Mulch it.* Mulch is a mixture of organic materials, such as tree bark, old wood chips, dead leaves, or compost. When placed around the base of a plant or tree, it stops moisture from evaporating, thereby reducing a plant's watering needs by up to 40 percent. Mulch also makes it difficult for weeds to grow and protects a plant's roots from the heat.

- *Use grass sparingly.* Because grass needs more water than any other type of plant in your yard, plant it only where it will be used as turf for walking or playing. If your grass is just for decoration, consider planting a lush ground cover instead (see Tip 30).

- *Mow better.* Your grass will be healthier, meaning you won't have to water it as much, if you do these five things:

  1. Always mow with a sharp blade.
  2. Mow only when your lawn is dry.
  3. Never cut the grass shorter than two inches.
  4. Change the mowing direction each time you mow.
  5. Cut off no more than one-third of the grass's current height.

- *Leave 'em on!* Start "grasscycling," that is, leaving your chopped grass clippings on your lawn instead of raking them off and bagging them. The clippings will decompose quickly and return precious nutrients to your lawn, allowing you to use at least 50 percent less chemical fertilizer a year. They'll also shade the soil and retain moisture, meaning you won't have to water your lawn nearly as much. And, because the clippings decompose so rapidly, they won't cause "thatch" (a lawn-killing layer of dead roots) to build up.

- *Go electric.* Consider replacing your carbon-dioxide-emitting, gasoline-powered lawn mower with a rechargeable electric one or even an old-fashioned push mower (a great way to exercise!).

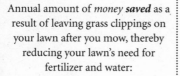

More Savings, Less Carbon Dioxide

Annual amount of *money* **saved** as a result of leaving grass clippings on your lawn after you mow, thereby reducing your lawn's need for fertilizer and water:

$24

Annual amount of CO_2 **not** *emitted* as a result of leaving grass clippings on your lawn after you mow, thereby reducing your lawn's need for fertilizer and water:

58 lbs.

Assumptions: Typical U.S. household uses two 47-pound bags of lawn fertilizer per summer (costing $18.82 each); grasscycling reduces fertilizer need by 50 percent; average U.S. household uses 84,738 gallons of water per year on outdoor activities, 32 percent of that on lawns; grasscycling reduces lawn water usage by 12 percent; cold water costs 0.15 cents per gallon; 1 gallon cold water saved = 0.018 pounds carbon dioxide not emitted.

Search For More Info

- www.ciwmb.ca.gov/Organics/GrassCycling/default.htm—Go here to learn more about the benefits of leaving grass clippings on your lawn after you mow.

- http://edis.ifas.ufl.edu/scripts/MG251—Check out this in-depth article on mulching.

- www.nativegrasses.com—This web site tells all about the grasses that are native to the United States.

Get Smart About Sprinklers

*Ninety percent of people overwater their
lawns and gardens.*

Overview. Designing a lawn and/or garden that needs minimal water
gets you only halfway. The other half of conserving irrigation water
lies in learning how to use your sprinkler properly. Overwatering
not only wastes water and energy but also is unhealthy for your
plants, since it flushes out the soil's rich nutrients. Overwatering also
promotes plant disease and weed growth, causes your plants to
become overcrowded, and damages their delicate root hairs.
Whether you use a complex drip-irrigation system or a simple hose,
a few simple changes can save you water, energy, and money off
your next water bill.

What You Should Know

- Adding an automatic rain sensor to your manual or automatic
  sprinkler is the *best* thing you can do to save water and cut
  your irrigation costs.

- The number-one factor that determines your sprinkler's efficiency is whether it sprays water *uniformly*. In many home

sprinkler systems, twice as much water needs to be sprayed to make up for uneven spraying.

- According to Florida's Water Management Department, if you water your lawn between 10:00 A.M. and 4:00 P.M., approximately 65 percent of the water will evaporate because of heat.

- Homeowners can save an average of 40 percent in irrigation costs simply by learning how to water their lawns properly.

- A typical garden hose without a nozzle flows at 12 gallons of water per minute!

- Most sprinkler systems today spray water five times faster than the soil can absorb it. As a result, most of the water is wasted through runoff. That's why you should water your lawn in short spurts (see table), with at least 30 minutes between spurts (applications).

Maximum Time a Sprinkler Should Be On per Application
(so the water has time to soak in)

Type of Soil Under Grass	Sprinkling Time Using a Spray Nozzle	Sprinkling Time Using a Rotor Nozzle
Sand	15–20 minutes	45–60 minutes
Loam	10–15 minutes	30–45 minutes
Clay	7–10 minutes	20–30 minutes

Adapted from "Water Management Options," North Carolina Department of Environment and Natural Resources.

- *Climate Results:* Attaching a rainfall sensor to your sprinkler can reduce your annual water consumption by up to 40 percent. If 100,000 people got rainfall sensors, they would prevent 30,000 *tons* of carbon dioxide from entering the atmosphere each year (thanks to all the energy saved by not having to pump as much water from the utility or well to their sprinklers).

- *Money Matters:* A rainwater collector installed on an average-size roof collects about 700 gallons of rainwater for every inch of rainfall. A typical household in Atlanta, Georgia (where it rains approximately 56 inches per year) can save up to $200 a year on water and sewage bills by collecting rain and using it instead of using utility water to water the lawn.

Easy Ways You Can Help

- *Put a nozzle on it.* If you have a regular old hose, think about attaching an inexpensive water-saving nozzle to the end of it. Look for one with an on-off switch and multiple spray options (mist, jet, and everything in between). Also, use a broom instead of a hose to clean off your driveway or sidewalk, and turn off the hose while you soap your car.

- *Assess your sprinkler.* Check your sprinkler's uniformity— if some plants are getting drenched and others are drying out, you may need to adjust the sprinkler's positioning. Make sure that you're not watering your sidewalk or driveway (not only does this waste water, it also contributes to runoff).

Last, be sure your sprinkler is watering plants at their roots, where they need the water to grow, not at their leaves.

- *Say when!* To avoid overwatering, don't turn your sprinkler on until you see the first signs of wilt (i.e., when your footprint stays in the grass after you walk on it). Memorize this: Plants grow best when they are watered slowly, deeply, and infrequently. Hmm, doesn't that sound a lot like . . . rain? Exactly. Plants have adapted to Mother Nature's habits. That's why you should water your lawn heavily once a week instead of lightly every day. Also, remember to:

  →Wait five to seven days after it rains before using your sprinkler.

  →Never use your sprinkler if there's a wind of more than 10 mph.

  →Water your lawn in the morning or evening, so the water doesn't evaporate in the heat.

  →Adjust your total sprinkling time every few weeks to go along with the changing seasons.

- *Get a timer.* To add a battery-operated timer to your sprinkler, simply screw it onto your faucet and then attach the hose. Timers can turn the water on for spurts of 15 minutes with 20 minutes in between so that the water has time to soak in. More advanced timers have settings for different days and weeks and can be set to activate only certain zones of sprinklers.

- *Add a rainfall sensor.* Adding a rainfall or soil-moisture sensor to your manual or automatic sprinkling system is the *best* thing you can do to save water. When either of these detectors senses that there's enough rain or moisture in the soil, it shuts off your sprinkler system and turns it back on again only when the moisture dissipates. For a cheaper (but not automatic) alternative, pick up a rain gauge or moisture-sensing probe at your hardware store and test the soil before you water your lawn or garden.

- *Do the drip.* If you're planning on upgrading your sprinkler system, consider getting a drip system. Replacing your sprinkler system with drip-emitting pipes that are placed low by the roots and spray heavy mists of water can reduce your irrigation costs by 60 percent. You have to be careful, though— if you don't take care of your system and adjust the timer according to the season, you could end up wasting twice as much water as you did before the upgrade.

- *Use recycled water.* If you really want to make an impact on the environment and eliminate nearly all your irrigation costs, make the extra effort to set up either a rainwater collector or a gray water system on your roof. A gray water system sends your soapy shower, dishwasher, and washing machine wastewater through a filter so that it can be safely reused on your landscape—providing your lawn with up to 100 gallons of water a day that would otherwise have gone down the drain! You could also place a rain-collecting barrel somewhere in your garden or at the bottom of a rain gutter and dunk in a watering can.

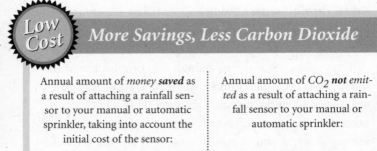

Low Cost

More Savings, Less Carbon Dioxide

Annual amount of *money* **saved** as a result of attaching a rainfall sensor to your manual or automatic sprinkler, taking into account the initial cost of the sensor:

$41

Annual amount of CO_2 **not** *emitted* as a result of attaching a rainfall sensor to your manual or automatic sprinkler:

599 lbs.

Assumptions: Average U.S. household uses 84,738 gallons of water per year outdoors; rainfall sensor reduces irrigation use by 40 percent; 1 gallon of cold water conserved = 0.011 kilowatt-hour conserved; cold tap water costs 0.15 cents per gallon; sensor costs $50, is self-installed, lasts five years.

Search For More Info

- www.sprinklerspot.com—Check out this web site to find almost any sprinkler part or product you can think of. Click on "Controllers" to see the selection of rainfall sensors.

- www.oasisdesign.net/greywater/greywater.htm—Go to this web site to learn more about gray water systems and how much water and money they can save you.

- www.watercasa.org/pubs/drip_top.html—Here are the top ten easy things you should do to make sure your drip irrigation system is performing efficiently.

Getting Around

Your Car: The Carbon-Dioxide-Spewing Monster

A typical lightweight vehicle in the United States emits 12,000 pounds of carbon dioxide a year and 54 tons over its lifetime—and there are 130 million lightweight vehicles in the United States.

Overview. Your car is responsible for emitting as much carbon dioxide a year as your entire house is. In fact, improving the fuel efficiency (miles per gallon) of your car is the single best thing you can do to prevent global warming. Why? Because gasoline- and diesel-burning automobiles are the largest single source of carbon dioxide emissions in the United States. In addition to carbon dioxide, they emit other pollutants, such as carbon monoxide and nitrogen oxide, which damage our crops and drinking water and cause smog, acid rain, respiratory problems, birth defects, and cancer.

So, what can you do to solve this problem? Three things. You can choose what *kind* of car you drive, how *often* you drive it, and *how* you drive it. This tip tells you *how* to drive your car so that it gets the best fuel efficiency possible. The next tip explains how you can find ways to drive your car less often, and Tip 39 describes what kind of car you should drive.

What You Should Know

- Gasoline is made from oil. The United States currently imports 54 percent of its oil, and two-thirds of the oil we use each day goes to transportation. By taking better care of your car and thereby improving its fuel efficiency, you will personally help reduce our country's dependence on foreign oil imports.

- The amount of carbon dioxide your car's tailpipe emits is directly proportional to the amount of gas your car uses. Furthermore, the vast majority of cars on the road today do not achieve their optimum fuel efficiency because their owners don't drive or take care of them as well as they could.

- Automobiles are the largest single source of carbon dioxide emissions in the United States, accounting for almost one-third of our total carbon dioxide emissions.

- Want proof that Americans love their cars? Every year for the past 10 years we have driven more miles than we did the year before.

- The world's population has doubled in the last 50 years, but the number of cars on the planet has grown *ten*fold. There are currently 530 million automobiles on Earth, which are responsible for 21 percent of the world's global-warming emissions.

- A tune-up will improve your car's fuel efficiency by about 15 percent. If your car is in really poor shape, a tune-up could improve its fuel efficiency by up to 50 percent!

- According to the Department of Transportation, one in four U.S. passenger cars have at least one substantially underinflated tire.

- A catalytic converter is a device in your car's engine that prevents most of the non–carbon dioxide air pollution (such as nitrogen oxide and carbon monoxide) from being emitted. It does *not*, however, stop carbon dioxide from being emitted, since there is currently no way to burn gasoline or diesel fuel without releasing carbon dioxide.

- *Climate Results:* If 100,000 people got their cars tuned up this year, they would save at least 124,000 *tons* of carbon dioxide from being emitted, thanks to improved fuel efficiency.

- *Money Matters:* If you always drive smoothly on the highway instead of accelerating and braking rapidly, you could save up to $120 a year in reduced gas consumption.

Easy Ways You Can Help

- *Keep 'er tuned.* A car that runs well uses less gas. Every month you should check to see if your radiator or anything else is leaking, if there's enough coolant in the radiator overflow bottle, if the battery terminals and air filters are clean, and if the drive belts are tight. You can check these things while you're filling up at the gas station. Refer to your owner's manual if you're not sure how to do any of these things.

Every spring and fall, you should change your car's oil and have a mechanic perform a basic tune-up—the improvement in fuel efficiency will more than make up for the cost of the tune-ups. Last, be aware of how often you head to the gas pump. If you notice you're suddenly filling up more often than usual, it probably means something is wrong with your car.

- **Easy there, tiger.** The more smoothly you drive, the less gas your car will use. Accelerating or braking rapidly when you're traveling at highway speeds can worsen your fuel efficiency by 33 percent! It's simple: The less you floor it and slam on the brakes, the less carbon dioxide will be emitted.

- **Don't idle.** Letting your car idle for just 20 seconds burns more gasoline (and therefore emits more carbon dioxide) than turning your car off and on again does! Furthermore, contrary to popular belief, idling can actually *harm* your car, since it causes gasoline to condense on and damage your cylinder walls and spark plugs. So, if you're ever going to be parked for more than 20 seconds, turn your engine *off*. The wear and tear on your ignition will cost you an average of only $10 a year, which will be repaid many times over by gas savings.

  Similarly, although it's tempting to warm up your car for 10 minutes in the winter, the resulting emissions are devastating to the environment and to your health. Studies have shown that driving is a much more effective way to warm your car up than idling is. And with today's electronic engines, you need to warm up your car for only 30 seconds on a winter day (to circulate the oil) before driving slowly away. For really cold climates, have a block heater installed for about $20. Put a timer on it so that it turns on an hour before you need to leave for work. Last but not least, don't use the remote-control starter if your car has one, since it increases the amount of time your car idles.

- *Slow down.* You can reduce your car's fuel consumption by 15 percent simply by driving 55 miles per hour instead of 65. That's because every car's fuel efficiency starts to plummet after 60 miles per hour. Plan ahead so you have plenty of time to get to your destination—then slow down, relax, and turn on some good tunes.

- *Inflate properly.* Keeping your tires properly inflated can reduce your gasoline consumption by 6 percent. That may not seem like a lot, but a 6 percent improvement in fuel efficiency will save the average car $40 a year in gas. More important, proper inflation *greatly* reduces the chances of having a blowout, which can cause everything from a simple flat tire to a fatal car accident. Always keep your tires at the maximum recommended pressure (check the label, usually located on your doorjamb, in the glove box, or inside the gas cap cover). Because tires lose pressure each month and with every 10°F drop in temperature, you should check your tires—including the spare—once a month. Buy a two-dollar pressure tester at your hardware store, keep it in the glove box, and check your tires while you're filling up.

- *And the list goes on!* There are many more things you can do to burn less gas while driving: Use the overdrive gear when you've reached a high enough speed, use cruise control, use high-octane fuel only if your owner's manual *specifically* calls for it, avoid heavy luggage to keep your car as light as possible, use the exact type of motor oil your owner's manual calls for, and use the air conditioner only when you're going over 40 miles per hour (otherwise open the windows).

No Cost

More Savings, Less Carbon Dioxide

Annual amount of *money **saved*** on gas as a result of always going 55 mph on the highway instead of 65:

$44

Annual amount of CO_2 ***not** emitted* as a result of always going 55 mph on the highway instead of 65:

720 lbs.

Assumptions: Typical car drives 12,000 miles per year, 4,800 of those on the highway, at U.S. average 23 miles per gallon; based on 2000 average price for regular unleaded gasoline of $1.41 per gallon; 1 gallon of gas burned emits 23 pounds carbon dioxide; driving 55 miles per hour instead of 65 reduces fuel consumption by 15 percent.

Search For More Info

- www.dnr.state.wi.us/org/caer/cea/projects/pollution/p2/ 2000/greener/Greener.pdf—Check out this easy-to-read brochure about how to treat your car and get the best fuel efficiency possible.

- http://oee.nrcan.gc.ca/autosmart/driving_main—Click on "A Guide on Maintenance" to learn more about the monthly checks and biannual tune-ups you should give your car.

- http://oee.nrcan.gc.ca/autosmart/idling/engine.cfm?Text=N —This Internet site explains exactly why idling can damage your car.

Learn How to Drive Less

Unless a store is one mile away or less, it's more energy-efficient to buy things on-line and have them delivered by truck than to drive to the store yourself.

Overview. Americans are addicted to their cars. We don't think twice before driving to pick something up at a store that's only three blocks away. In fact, we romanticize driving with our thoughts of long Sunday drives and adventurous cross-country road trips. Well, we need to get rid of this addiction—and fast—because driving is harmful to *our* health and to our *planet's* health. The air pollutants that spew out from our tailpipes have been proven to cause cancer, and the carbon dioxide that's emitted causes global warming. One of the easiest ways to reduce these emissions is simply to change our mind-set about driving. We need to make driving our *last* choice of transportation, not our automatic first. And the first step in achieving this new mind-set is to reorganize our lives so we don't need to drive as much, or as far.

What You Should Know

- Studies have shown that telecommuters who work from home are happier and at least 30 percent more productive in their jobs than their co-workers are back at the office.

- In 2001, 30 million Americans worked from home at least one day a week instead of at the office. That's 20 percent of the working population. At least 40 million Americans are expected to be telecommuting at least one day a week by 2004.

- Buying a gift for someone on-line and having it shipped via ground transport directly to the recipient saves more energy than any other type of on-line purchase.

- The most important decision you can make after you decide to do your shopping on-line is how to ship your purchases. Even though planes and trucks emit much more carbon dioxide than cars do, because they carry hundreds to thousands of packages at a time, they end up being more energy-efficient than you would be driving to and from the mall. This table compares the amount of gasoline used per item with different purchasing and shipping methods:

Mode of Travel	Amount of Gas Used Per Item
Driving 20 miles round trip to a mall	1 gallon
Shipping 1,000 miles via air freight	0.166 gallon
Shipping 1,000 miles via truck	0.100 gallon

Adapted from http://www.cool-companies.org/energy/5.cfm

- Doing multiple errands in one trip will save you gas because you'll be driving fewer miles, but you'll save even more gas if each errand takes 20 minutes or less. That's because a warm engine is much more energy-efficient than a cold engine.

- *Climate Results:* If just 20,000 people worked at home one day a week instead of driving to the office, they would prevent at least 15,000 *tons* of carbon dioxide from being emitted each year. Plus, they would help reduce rush-hour traffic.

- *Money Matters:* If you order your groceries on-line and have them delivered to your home each week, you'll save at least $20 a year on gas, 35 hours of shopping time, and 10 hours of driving time.

Easy Ways You Can Help

- *Support local businesses.* Try to do your errands closer to home to cut down on the amount of driving you do. By shopping at local stores and eating at nearby restaurants, you'll also be supporting your local economy!

- *Do two things at once.* Postpone errands until you can combine two or more into one multistop trip. Then, try not to retrace your route and, if you can, park your car at one stop and walk to your other errands. Besides saving time and gasoline, you'll also reduce wear and tear on your car.

- *Shop on-line!* Even though the products you order on-line are delivered to you via large, carbon-dioxide-emitting trucks (or trains or planes), it's still 40 percent more energy-efficient to buy products on-line and have them shipped via air freight rather than drive to the store yourself—and 90 percent more efficient to have them shipped via truck. That's because trucks and planes deliver many packages at once, which increases their efficiency.

Try to plan ahead so you can always choose the slowest method of shipping (which saves the most energy). When ordering gifts on-line, always see if you can ship directly to the recipient. Banking and paying your bills on-line also saves energy because less paper and ink need to be manufactured and less mail needs to be delivered to you by big trucks. We should note that shopping on-line does not stimulate your local economy. Try to achieve a balance—support the stores closest to you and shop on-line for everything else.

- *Say good-bye to video rentals.* Consider getting a pay-per-view cable box, which allows you to view the most recently released movies directly from your TV for about four dollars each. Since 80 percent of movie rentals today are new releases, you'll get to see the movies you want. Plus, you'll save gas (and emissions) by not driving to and from the video store to pick up the movie and again to return it. You also won't have to worry about late fees! Call your cable company today to ask about installing a pay-per-view box. Save the drive to the video store for those rare times when you're in the mood for an "oldie but goodie."

- *Work from home more often.* Ask your boss if telecommuting even one day a week is an option. Thanks to the advent of e-mail and video conferencing, communicating with your office and clients while you work at home is no longer an issue. Telecommuting not only eliminates the pollution your car would have emitted but also helps alleviate the extra pollution that's emitted when traffic jams occur.

- *Come closer.* The next time you're planning to move, try to move closer to your workplace. You'll be much happier with the commute, and the Earth will be much happier with you. Also, consider moving to a more compact community so your car trips will be shorter and your family will be able to walk or bike to school, the grocery store, or even the movies.

- *If you must drive . . .* and you own two cars, always take the one that gets better fuel efficiency.

No Cost   *More Savings, Less Carbon Dioxide*

Annual amount of *money* **saved** as a result of combining four 6-mile-round-trip errands each week into one 10-mile round-trip, multistop errand:

$45

Annual amount of CO_2 **not** *emitted* as a result of combine four 6-mile-round-trip errands each week into one 10-mile-round-trip, multistop errand:

728 lbs.

Assumptions: Based on 2000 U.S. average price for regular unleaded gasoline of $1.41 per gallon, average U.S. passenger car gets 23 miles per gallon; 1 gallon gasoline burned = 23 pounds carbon dioxide.

Search For More Info

- www.cool-companies.org/energy—Check out this wonderful web site to learn how much energy on-line shopping saves compared with driving to the mall.

- www.workingfromanywhere.org and www.gilgordon.com— These two web sites offer support, advice, and statistics about telecommuting in the United States.

- www.peapod.com and www.netgrocer.com—These are two of the largest on-line grocers. Many major grocery stores (such as Jewel and Albertson's) also offer on-line order and delivery options through their web sites—check to see if yours does.

Let Someone Else Do the Driving

Car pooling to work with one other person instead of driving by yourself prevents approximately 8,520 pounds of carbon dioxide from being emitted each year.

Overview. Aren't you tired of traffic jams? Wouldn't you rather get in some extra winks of sleep or read a book than drive to work every day? Well, it's as simple as taking public transportation or car pooling. In the last tip, you learned how to reorganize your life so you don't need to drive as much. That won't completely kick our addiction to driving, though. We also need to remember that there are alternatives to driving. For longer trips, we can take public transportation or carpool; for shorter errands, we can bike or walk. These alternatives are cheaper *and* help prevent global warming.

What You Should Know

- Car pooling saves you gas money and prevents carbon dioxide from being emitted, but it also leads to less wear and tear on your car—which means fewer maintenance costs.

- Over the next five years, the U.S. government plans to spend $175 billion on highway transportation, and only $4.2 billion on public transportation. Imagine how much faster, cleaner, safer, and far reaching our public transportation systems would be if those numbers were reversed!

- Many business owners are cashing out their employee parking spots to encourage car pooling and the use of public transportation. Cashing out is when employers give their employees monthly travel stipends but then charge them for parking spots. As a result, many employees decide to save money by car pooling or taking less expensive public transportation.

- *Climate Results:* If just 500 people took public transportation to work one day a week instead of driving, they would prevent approximately 300 *tons* of carbon dioxide from being emitted each year.

- *Money Matters:* Commuting into a city by car costs from $2,500 to $6,000 a year for the permanent parking space, plus an additional $500 or so in gasoline costs. By contrast, taking a commuter train costs roughly $1,600 a year, while commuting on the bus or subway costs only $750 a year.

Easy Ways You Can Help

- *Take public transportation.* One full city bus means about 40 fewer cars on the road. The environmental benefits of public transportation are enormous, but there are personal benefits too: you won't have to pay for gasoline or parking, you won't have to drive in traffic jams, and you can read, get work done, listen to music, or even sleep during your commute. Check out your city's official web site, call your local government, or select your state and city on this web site—www.apta.com/sites/transus—to find out about public transportation options in your area.

- **Fight for better public transportation.** Are your city's buses slow or unreliable? Is your city's subway unsafe? If these things are keeping you from using public transportation, speak up! Governments *will* listen to people who actually take a few minutes to make a phone call. They'll listen even more if you get some people together and voice your concerns at the next city council meeting. Call your local government today or ask the American Public Transportation Association (www.apta.com) to help you improve your city's public transportation system.

- **C'mon and carpool already!** Car pooling is fun (you get to be social before you start your workday), fast (the Carpool Only Lane lets you fly by everyone else), saves you gas money, and most important, takes carbon-dioxide-emitting cars off the roads. Car pooling isn't just for commuters, either. You can carpool with other parents to take your kids to school, carpool with friends to go out to dinner, and more. To find someone with whom to carpool, check out the web sites in "Search for More Info."

- **Get money for just sitting there.** Ask if your company gives rebates, benefits, or incentives for commuting by public transport or car pool, such as subsidized transit passes or free parking for car poolers. If they don't, ask them to! Many companies also offer free cab rides home in case of an emergency and a free shuttle bus to favorite lunch spots.

- **Carshare.** Welcome to the transportation of the future: car sharing. It started in Europe about 15 years ago, and it's already arrived in 11 U.S. cities. Car sharing allows you to enjoy the benefits of having your own car without having to foot the whole bill. Here's how it works: When you need a car, whether for four hours or four days, you simply call and reserve the kind you need (car, van, SUV). Then pick it up at one of many convenient locations around your city, drive it where you need to go, and return it when you're done.

  In addition to a nominal monthly fee that covers insurance and maintenance costs, you pay only for the gas you use. Car sharing gives people who can't afford a car on their own the opportunity to drive. If you currently own a car but drive fewer than 7,500 miles a year, joining a car-sharing network will save you money. More important, car-sharing makes people realize that they can fulfill most of their driving needs with public transportation. As a result, people who switch from owning their own car to joining a car-sharing network usually reduce their total car-driving time by 50 percent. Talk about a reduction in carbon dioxide emissions!

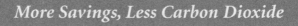

More Savings, Less Carbon Dioxide

No Cost

Annual amount of *money **saved*** as a result of car pooling one day a week instead of driving, assuming you'll split the cost of gas with your fellow car pooler:

$33

Annual amount of CO_2 ***not*** *emitted* as a result of car pooling one day a week instead of driving:

543 lbs.

Assumptions: Based on a 10-mile one-way (20 miles round trip) commute; both cars get U.S. national average 23 miles per gallon; 2000 U.S. national average retail price of $1.41 per gallon regular unleaded gas; 1 gallon of gas burned = 23 pounds carbon dioxide emitted.

Search For More Info

- www.carpoolconnect.com/states—Go to this web site to find other people in your area who want to commute to work, share the expenses of a long road trip, or carpool to a special event. Many cities' public transportation web sites list car-pooling information as well.

- www.carsharing.net—Scroll to the bottom of this page to find out which U.S. cities have car-sharing networks. Click on the "Library" link to learn how to start a car-sharing program in your city.

- www.fta.dot.gov/library/policy/cc/cc.html—These links explain the rewards employers can offer their employees for commuting to work via public transportation or car pool instead of by car.

Use Person Power

One-quarter of all automobile trips in the United States are less than one mile long.

Overview. What could be more American than going for a Sunday drive? That's just the problem. Americans don't realize how much our big cars are hurting the environment. Did you ever wonder why Europeans drive such small cars? It's not so much because their cities are overcrowded but rather because their gas prices are so high—and smaller cars get better fuel efficiency. We may be fortunate enough to enjoy lower gas prices and afford bigger cars, but that doesn't mean we have the right to harm a planet we share with six billion other people. And there's no question that our cars and trucks are hurting the environment—they account for nearly one-fourth of the United States' total greenhouse gas emissions, making them our biggest single source of emissions. So what can we do? Well, for starters, how about going for a Sunday bike ride?

What You Should Know

- Walking or biking instead of driving reduces air pollution and saves you gas money. The benefits don't stop there, though. These forms of transportation are also quieter, safer, and cleaner. They lower our dependence on foreign oil imports, provide exercise, reduce stress, contribute positively to the economy, have a minimal impact on the natural landscape, and are more social than driving (for example, you can stop on the sidewalk to chat with someone). Talk about benefits!

- Walking and biking are both statistically much safer than driving in a car. In fact, 86 percent more people die in car accidents each year than they do from walking. Furthermore, 50 times as many people are killed in car accidents each year than they are in biking accidents—and 96 percent of those who died in biking accidents died because they weren't wearing helmets.

- All new commercial developments in Los Angeles are required to include bike parking, showers, and lockers. Now that's planning for an environmentally friendly future!

- The electric bike industry is one of the fastest-growing transportation markets in the world. That's because it has so many applications—tourists can use electric bikes to spend a day sightseeing without getting tired, police departments can use them to patrol the streets for longer shifts, and senior citizens can use them to get around more easily.

- The Centers for Disease Control recently reported that more than half of U.S. adults are considered overweight. Meanwhile, bicycling burns approximately 400 calories per hour.

Plus, if you incorporate walking or biking into your life, you can say good-bye to that hefty gym membership fee—both biking and walking are outdoor, lifelong forms of exercise that are not only free but also easy on your joints.

- Most people know that biking or walking instead of driving reduces air pollution. But did you know that they also reduce water pollution? Every year hundreds of thousands of gallons of oil, antifreeze, and brake fluid drip from our cars and get washed into our waterways, eventually making their way into lakes and rivers, where they pollute our drinking water and poison our fish and wildlife. The only thing you drip by walking or biking is sweat!

- *Climate Results:* If you use an electric bike to commute five miles each way, one day a week instead of driving to work, you'll prevent 515 pounds of carbon dioxide from being emitted every year.

- *Money Matters:* You already know that walking or biking instead of driving means significant gasoline savings. Don't forget that you also won't need to pay as much for parking, bridge tolls, and maintenance. If you go completely without a car, you also save on insurance, registration, and licenses.

Easy Ways You Can Help

- *Go for a walk.* Walking is a zero-emissions mode of transportation, meaning it's completely pollution-free. Plus, it doesn't cost you a thing. If you can walk instead of drive somewhere, do it. Go for a romantic evening stroll instead of driving to the movies, have the kids walk to school if possible, and do errands on foot if the stores are close enough. Basically, if the weather is nice and the trip is under a mile, walk! You'll be helping your body, your wallet, *and* the environment.

- *Dust off your bike.* Forty percent of all automobile trips in the United States are two miles or less in length—that's perfect for a bike ride! And with today's larger and sturdier cargo baskets (or pull-behind trailers), doing your errands by bike is easier than ever. Besides being environmentally friendly, bicycling also reduces stress (through exercise) and allows children and senior citizens to be more mobile.

- *Commute by bike.* Can you bike to work? Over five million Americans do! Just think—you'd get your workout out of the way, you'd never have to worry about traffic jams or parking spots, and you'd save hundreds, if not thousands, of dollars each year in gasoline and parking costs. Not only will you look forward to your commute, but it will be the highlight of your day. Even biking to work one day a week will reduce your commuting costs by 20 percent. If it's really too far to bike to work, leave a bicycle at your office so you can use it to run errands during lunch hour.

- *Bike without breaking a sweat.* If you're excited about biking to help the environment but not about showing up at your destination drenched in sweat, consider a rechargeable electric bike. These are just like regular bikes but with small, battery-powered motors to give extra power. Going uphill or into the wind is as easy as cruising on a flat surface. You'll be able to take longer trips and carry heavier loads, such as groceries. You can let the motor provide all the power you need to ride the bike or you can let it provide only as much power as you want (by turning the handle grip proportionately). You can also turn the motor off completely and ride it like a regular bike.

Even though electric bikes use electricity, which is produced by the burning of fossil fuels at power plants, they emit 99.5 percent less carbon dioxide than gasoline-fueled cars. To be completely emission-free, buy an electric bike whose batteries are recharged using wind or solar power (see Tip 48). Also, check if your state offers tax incentives or rebates for buying an electric bike by calling its energy office (go here to find the number: www.naseo.org/members/states.htm).

- *Travel electrically*. Electric bikes are only one type of lightweight electric vehicle (LEV). Electric scooters and one-person electric cars are becoming increasingly popular with commuters. Although they use more electricity than electric bikes do, they still emit much less carbon dioxide than a car. See "Search for More Info."

- *If you must take a taxi . . .* take a pedicab! Many big cities, such as New York City, San Francisco, Denver, and San Diego, now have electric bikes pulling semicovered passenger trailers. If walking to your destination is not an option, pedicabs are much more fun and environmentally friendly than taxis.

- **Lobby for better sidewalks and bike lanes.** The organizations listed in "Search for More Info" can help you and your fellow community members learn how to lobby for wider sidewalks, better bike lanes, bike racks on city buses, and everything else you'll need to make walking and biking safer and easier for your community.

More Savings, Less Carbon Dioxide

Annual amount of *money* **saved** as a result of taking three 8-mile-round-trip bike rides each month instead of driving to your destination:	Annual amount of *calories* **burned** as a result of taking three 8-mile-round-trip bike rides each month instead of driving to your destination:	Annual amount of CO_2 **not** *emitted* as a result of taking three 8-mile-round-trip bike rides each month instead of driving to your destination:
$18	**11,520**	**288 lbs.**

Assumptions: Based on U.S. 2000 average regular unleaded gas price of $1.41 per gallon; average U.S. passenger car gets 23 miles per gallon; 1 gallon gas burned = 23 pounds carbon dioxide emitted; biking burns 400 calories per hour when biking 10 miles per hour; assumes the person already has a bicycle.

Search For More Info

- www.americawalks.org and www.bikeleague.org—These two organizations offer advice on how to lobby for better pedestrian and biking conditions respectively in your community.

- www.electric-bikes.com/lev.htm—Go here to see all the types of lightweight electric vehicles on the market today, including electric bicycles, pedicabs, scooters, one-person commuting cars, and more.

- www.shophunter.com/Sports_and_Fitness/Cycling—Go here for links to the top 10 places to buy bicycles on-line.

Trains, Not Planes

Flying from Chicago to New York emits six times as much carbon dioxide (per person) as taking a train, and 16 times as much carbon dioxide as taking a bus.

Overview. So far, we've been telling you some of the best ways you can prevent global warming. Well, now it's time for us to tell you the single worst thing you can do to *contribute* to global warming: fly on a plane. Planes are like giant, dirty diesel trucks cruising through the sky. Buying a new gas-electric hybrid car may reduce your carbon dioxide emissions by 6,341 pounds a year (see Tip 39), but just one round-trip flight from New York to San Francisco will wipe those savings away, since the plane for that trip will spew out 6,450 pounds of carbon dioxide *per passenger!* Flying is different from driving, though. We don't expect you to go out and buy yourself a fuel-efficient plane anytime soon. However, there are alternatives to flying that are cheaper and emit much less carbon dioxide. Plus, with a bit more funding from the government, our trains just might get you there faster than a 747.

What You Should Know

- Airplane travel is responsible for roughly 8 percent of the world's global warming emissions. That's not as much as passenger vehicles or home electricity use, because only a small percentage of people fly on planes compared with the number of people who drive cars or own homes. Those people who do fly regularly, however, are contributing *more* than their share of carbon dioxide emissions.

- Each year 600 million Americans fly a total of 70 billion miles—that's 107 million tons of carbon dioxide emitted each year!

- In 2001, a record-breaking 23.5 million passengers rode Amtrak, our nation's interstate railway company, and the numbers keep growing.

- *Climate Results:* If 100,000 people took the new high-speed Acela train from New York to Boston instead of flying, they would collectively prevent 7 *million pounds* of carbon dioxide from being emitted.

- *Money Matters:* The United Kingdom plans to spend $21.5 billion on railway transit over the next five years. For the same number of years, the United States plans to spend only $2 billion on railways, even though we have a population almost five times the size of the United Kingdom's.

Easy Ways You Can Help

- *Catch a Greyhound instead.* Try to plan your travels ahead so that you can give yourself time to take a bus—the cheapest and least polluting way to travel long distances. Almost all today's intercity buses have reclining seats and show movies while you ride. Check out Greyhound, the nation's largest interstate bus line, to see if you can make your next trip by bus (www.greyhound.com).

- *Take the train instead.* Faster than the bus and at least twice as energy-efficient as plane travel, intercity trains are becoming more and more popular. With spacious sleeper cars and sit-down dining cars, traveling by train is romantic and affordable, even if it does take a bit longer. Check out Amtrak's routes here: www.amtrak.com.

- *Ride the fast train.* Bullet or high-speed trains have been zipping across Europe and Japan at speeds of 200 miles per hour for the last decade. In the fall of 2000, the United States launched its first high-speed train, with a top speed of 150 mph, between Boston, New York City, and Washington, D.C. Called the Acela, this train gets you from Boston to NYC in three hours and 23 minutes, and from NYC to Washington in two hours and 44 minutes, costing about the same as it would to fly and take cabs to and from airports. More high-speed trains are planned for California, the Texas-Oklahoma-Arkansas area, the Pacific Northwest, the Chicago–Great Lakes area, the Gulf Coast, and the entire East Coast.

- *Get ready for maglev.* A train that runs on magnetic levitation is the perfect combination of environmental consciousness and speed. These trains, which are propelled by magnetic force, travel at 300 miles per hour. Maglev trains will greatly reduce global warming emissions and air pollution as well as create jobs, lessen airport and highway traffic, and decrease our dependence on foreign oil imports. Germany and Japan are both testing out maglevs that should be ready for public use in a few years.

  The U.S. government recently held a competition to determine the best place to build our first maglev train—the finalists are Baltimore–Washington, D.C., and the Pittsburgh area. After a little more research and planning, the government will award $950 million to the more promising of the two projects. The winning city may have the United States' first maglev train up and running by the end of the decade. Other maglev railways are in the works for California, Florida, Georgia, Louisiana, and Nevada.

- *Road trip it.* If time permits, go for the long haul and drive. You'll emit less carbon dioxide than you would by flying, no matter how long or short your flight is. And you'll have more flexibility once you get to your destination. If four people ride in the car, it's typically less polluting than taking a train, but if three people are in the car, it's better to take the train. Also, do some math and see if renting a car that gets excellent fuel efficiency (such as a gas-electric hybrid, see Tip 39) would be cheaper than taking your own car.

- ***If you must fly . . .*** buy carbon offsets to compensate for the global warming emissions your flight will cause. First, use this web site—www.chooseclimate.org/flying/mapcalc.html —to figure out how many pounds of carbon dioxide per passenger your flight will emit (to convert kilograms of carbon dioxide into pounds, multiply by 2.2). Then go to the web site of a carbon-offset company, such as Climate Partners or TripleE (see "Search for More Info"), and calculate how much money you should send them to make up for your flight's carbon dioxide emissions. These companies will invest your money in emission-reducing projects such as wind power plants, car-pooling organizations, and tree-planting programs. By buying carbon offsets, you balance out the carbon dioxide emissions you caused by flying.

No Cost *More Savings, Less Carbon Dioxide*

Amount of *money* **saved** as a result of taking a round-trip bus for 450 miles each way (the distance from Boston to Washington, D.C., or from L.A. to San Francisco) instead of flying:

$50

Amount of CO_2 **not** *emitted* as a result of taking a round-trip bus for 450 miles each way (the distance from Boston to Washington, D.C., or from L.A. to San Francisco) instead of flying:

3,346 lbs.

Assumptions: Flying 450 miles emits 1,763 pounds of carbon dioxide per passenger; a 40-passenger bus emits one ton of carbon dioxide per 10,000 miles; round-trip bus ticket from Boston to Washington, D.C., costs $70; round-trip plane ticket costs $120.

Search For More Info

- Check out these two carbon-offset companies: Climate Partners (www.climatepartners.com), and TripleE (www.tripleE.com).

- www.fra.dot.gov/rdv/hsgt—Go here to learn about what's going on with high-speed trains in the United States today. Also, check out this map of approved and pending high-speed rail corridors across the country: www.fra.dot.gov/rdv/hsgt/states.

- www.amtrak.com/trains/acelaexpress.html—Check out the Acela, the new high-speed train that connects Boston, New York, and Washington, D.C.

New Cars, New Fuels

*Minivans, SUV's, and pickup trucks make up
50 percent of all U.S. vehicles. These vehicles get an
average of 18 miles per gallon of gas, compared
to a typical passenger car's fuel efficiency of
24 miles per gallon.*

Overview. This is a long tip, but that's because it's the most important one in this book. Replacing your current car with a fuel-efficient one is hands down the *best* thing you can do to save the most carbon dioxide and money in one move. Obviously, not everyone has the money to buy a new car this year. But when you do, we want you to remember this tip. Why? Because buying a hybrid car that gets excellent fuel efficiency or, better yet, an alternative-fuel car that runs on a renewable fuel, does more than help prevent global warming and save you gas money. It also lessens our dependence on foreign oil imports; reduces air pollution, smog, and acid rain; reduces the chances of oil spills; and, perhaps most important, sends a message to car manufacturers that Americans want cars that don't harm the environment.

What You Should Know

- By taking fuel efficiency into consideration when you buy your next car, you will personally help reduce our country's dependence on foreign oil in addition to reducing your car's global warming and air pollution emissions. Ultimately, using a fuel that is renewable and that we can produce domestically (such as ethanol, which can be made from corn or switch-grass) is the only thing that will *completely* free us from our dependence on foreign oil.

- The U.S. national fuel economy levels were at their best 20 years ago. Because of the ensuing popularity of gas-guzzling SUV's and minivans, however, our national fuel economy levels have since dropped and are still lower than they were in 1988.

- Because trucks are traditionally heavier and bigger than passenger cars, they do not have to meet fuel-efficiency standards as strict as those of lighter passenger cars. Unfortunately, SUV's and minivans are classified as light-duty trucks (even though they are used as passenger cars), which is why our government *allows* them to have such horrible fuel efficiency.

- Even though fossil fuels are burned both in your car's engine and at power plants, your car's tailpipe pollution is more dangerous to your health than power plant pollution because of its proximity—you breathe the fumes in every time you're near a vehicle.

- **Climate Results:** If you buy a new car that gets only 3 more miles per gallon than your old car, you'll save 30,000 pounds of carbon dioxide over the lifetime of your new car.

- **Money Matters:** Recent breakthroughs in aluminum technology allow cars to be built 40 percent lighter than ones built with steel, while testing *safer* in crash tests. For every 10 percent decrease in vehicle weight, your fuel efficiency improves by 7 percent. So, buying a car with this new aluminum siding will save you more than $1,500 over the life of the car.

Easy Ways You Can Help

- **Fuel efficiency, fuel efficiency, fuel efficiency!** We can't say it enough—this is *the* most important factor to consider when buying your next car, in terms of saving money and the planet. It's plain and simple: the more gas your car uses, the more carbon dioxide it emits and the more you will contribute to global warming. The easy solution? Buy the car that gets the best fuel efficiency (the most miles per gallon of gas) and also fits your needs. The two web sites listed under "Use These Web Sites to Help Buy Your Next Car" can help you find the most fuel-efficient kind of car you're looking for.

- **Hurray for hybrids!** In the words of Toyota Prius customers, Americans are saying "Yes!" to hybrids. And it's no wonder—with an average of 46 miles per gallon of gas (that's the new, adjusted EPA rating, which takes into account air-conditioning use, faster speeds, etc.), and with gas prices soaring above $3 per gallon in many parts of the country, the Prius is a sweet solution to high gas prices and global warming. In addition, it also offers an affordable price tag, outstanding reliability, and chart-topping owner satisfaction ratings.

After the success of the Prius, many other automakers jumped on board and released hybrids—even SUV hybrids. For some of these, though, a small increase in fuel efficiency did not make up for the price that carmakers charged for them. Our advice is simple: If you're going to buy a hybrid (and we hope you do), buy the one with the highest fuel efficiency for its class. For example, if you must drive an SUV, consider the latest hybrid models that get upward of 32 miles per gallon of gas, like the Ford Escape Hybrid.

Some basics: Hybrids *never* need to be plugged in. Just fill up your hybrid at a gas station and then enjoy the freedom of not filling up again for 500 miles (in the case of the Prius). Some hints of the future: Technology is moving fast, and we may very well see hybrids that get more than 90 miles per gallon of gas within five years. Don't let that stop you, though—if you're ready to buy a new car now, buy the most fuel-efficient hybrid you can. If a 90-mpg hybrid appears in five years, you can trade your old hybrid in for the new, while enjoying the high resale values of hybrid cars.

- ***Buy an electric (or plug-in hybrid) car.*** Electricity is considered an alternative fuel because it releases no emissions whatsoever when used to power a car. However, since fossil fuels are burned when the electricity is created at the power plant, these cars are not completely emission-free (unless, of course, the electricity comes from a renewable source, such as solar or wind power—see Tip 48). Still, even taking into account the fossil fuels burned at a coal power plant, electric vehicles emit 67 percent less carbon dioxide than do gasoline-powered cars, and about 17 percent less carbon dioxide than today's best hybrids. Last, tests have proven that electric vehicles are just as safe as regular cars.

  Until recently, electric vehicles (EVs) could not travel very far or very fast. Thanks to recent breakthroughs in battery technology, however, EVs such as the Tesla Roadster can travel 200 miles between battery charges and go as fast as 130 miles per hour. The Tesla is a sportscar, though. Other EVs coming to market will have a more affordable price tag and will be able to travel around 40–50 miles between charges (perfect for daily use and commutes). Also coming to market soon are plug-in hybrid vehicles, which are completely electric for the first 30 miles or so, and then switch over to the hybrid engine to allow you to drive an additional 500 miles before filling up with more gas.

  Recharging is easy: Simply plug in your car overnight and pay off-peak prices for the electricity. Eventually, there will be electric charging stations at all gas stations, which will allow you to travel cross-country with ease. Also, electric motors are three times more efficient than gasoline ones, which is why owning an electric car is 40 percent less expensive than owning a gasoline-powered car (taking into account the cost of the electricity versus the cost of gas). Need another reason to go electric? Many states provide significant tax rebates for buying an EV.

- *Corn—the next gasoline?* Gasoline is dirty. Diesel is dirtier (see page 243 to find out why). The good news: There are cleaner, less polluting fuels out there that you should know about because, chances are, they will soon replace gasoline as we know it. Ethanol, typically made from corn but more cleanly made from switchgrass, is the most widely used alternative fuel. Because it is made from plants, ethanol is renewable (unlike fossil fuels), and it has no *net* carbon dioxide emissions. That's because the plant soaks up as much carbon dioxide when it's growing as the ethanol emits when it's burned in a car's engine (see Tip 31). However, emissions *are* produced in the planting, harvesting, and processing of ethanol, which is why it's so important to practice sustainable farming. One final perk: Ethanol is domestic; oil is (mostly) foreign.

  Almost all modern cars can already run on gasoline that has been blended with a little bit of ethanol, making it a slightly cleaner fuel. This blend, called E10, is becoming more widely available at gas stations across the nation—check your owner's manual to see if *your* car can run on E10.

- *Fuel cells of the future.* Some say that hydrogen fuel cells will completely replace internal combustion engines (the engines that currently power our cars) within 15 years. A hydrogen fuel cell is a device that, through a chemical reaction, converts hydrogen *directly* into electricity, without going through combustion. Fuel cells are two to three times more efficient than the combustion engines we have now, but best of all, the only by-products of the chemical reaction are water and heat, which means *zero* global warming emissions.

As with electric vehicles, however, there's a catch. Although a fuel cell is emission-free, obtaining the hydrogen that a fuel cell needs in order to work can cause global warming emissions or air pollution, depending on how you obtain it. And even though hydrogen is the most abundant element on Earth, it hardly ever appears on its own. It's usually attached to another element (for example, in water, hydrogen is attached to oxygen), and it takes energy to separate these elements.

Scientists are exploring many ways in which they can obtain hydrogen in an eco-friendly way. The first generation of hydrogen fuel cell cars will probably obtain their hydrogen by burning fuels like methanol. The actual fuel cell will make the car very efficient (around 80 miles to the gallon!), but because it will need to burn methanol in order to separate out the hydrogen, some carbon dioxide and air-polluting gases will still be emitted. In 10 to 20 years, however, we will most likely get our hydrogen from renewable sources such as water or even algae— and those cars will be *completely* emission-free!

The technology for a hydrogen fuel cell that gets its hydrogen from methanol is *already here.* In fact, all of the major automobile manufacturers are showing off their concept fuel cell cars at exhibitions. The only reason they're not on the market yet is because fuel cell technology is still expensive. As soon as scientists find a way to make this technology more affordable, these cars will be giving electric vehicles a run for their money.

- **Don't be tricked by diesel.** Cars that run on diesel fuel usually get better fuel efficiency than similar, gasoline-powered cars, but don't think you're helping the environment by buying a diesel-fueled car! A catalytic converter—the thing in your engine that reduces your tailpipe pollution—doesn't work as well with diesel fuel. Also, diesel exhaust is *10 times* more carcinogenic than gasoline. Although diesel technology has gotten cleaner in the past few years, diesel still comes from fossil fuels. If you're looking to save money at the pump, buy a hybrid instead of a diesel.

- **Get a flexible- or dual-fuel tank.** A flexible-fuel car has a gas tank that can accept either pure gasoline, pure methanol or ethanol, or gasoline mixed with any amount of ethanol or methanol. Methanol is made from either natural gas (a fossil fuel, which is bad) or wood (a renewable source, which is good). Many new cars are being built with these special gas tanks. Meanwhile, a dual- or bifuel tank is exactly what it sounds like: a car that has two separate fuel tanks—one for gasoline, and one for either natural gas or propane. The car switches automatically between the fuel tanks, depending on which one has fuel in it. To get a dual-fuel tank, you can either buy a new car with a dual-fuel tank or add a second gas tank to your current car.

Natural gas (also known as compressed natural gas—CNG) and propane (also known as liquid petroleum gas—LPG) are both fossil fuels, but they're considered alternative fuels since they burn so much cleaner than gasoline. They emit 20 to 50 percent less global warming–causing pollutants than gasoline. They're also much cheaper than gas. Generally speaking, natural gas is cleaner than propane, but propane is currently the most accessible of all alternative fuels (there are propane fueling stations in all 50 states).

- *Lighter is better.* A heavier car needs more gas than a lighter car, so to cut down on your carbon dioxide emissions *and* save gas money, buy the lightest car you can. Some people think that buying a heavier car is safer; however, air bags, seat belts, and a safe structural design (one that doesn't roll over easily, as many SUV's do) all increase your chances of survival *much* more than additional weight does. Also, thanks to advances in lightweight aluminum technology, buying a lighter car no longer means you have to sacrifice safety. For example, the aluminum body of the Acura NSX sports car is 40 percent lighter but has *higher* rigidity and crashworthiness than a sports car made of steel.

- *Natural gas: a good transition fuel.* So far we've mentioned cars that use gasoline *and* an alternative fuel, but you can also buy cars that run *only* on alternative fuel, such as natural gas or propane. These cars are less common since the fuel isn't yet as widely available as gas is, but they are *the* cleanest cars on the market in terms of reducing both global warming emissions and air pollution—95 percent cleaner than gasoline-powered

Investment

More Savings, Less Carbon Dioxide

Annual amount of *money **saved*** as a result of buying a gas-electric hybrid car instead of a typical 24 mpg gaso-line-powered car:	Annual amount of CO_2 ***not*** *emitted* as a result of buying a gas-electric hybrid car instead of a typical 24 mpg gaso-line-powered car:	Total amount of *money **saved*** as a result of buying a gas-electric hybrid car instead of a typical 24 mpg gaso-line-powered car:
$543/year	**5,497** lbs./year	**$4,883**/9 years

Assumptions: Both cars cost $21,000 and last nine years; hybrid and passenger cars get 46 and 24 miles per gallon, respectively; U.S. average 12,000 miles driven per passenger car per year, 2005 U.S. average retail price of $2.27 per gallon unleaded gas; 1 gallon gas = 23 pounds carbon dioxide emitted.

cars! Canada has one of the world's largest supplies of natural gas, so switching to natural gas vehicles offers the United States a way to reduce its dependence on foreign oil. Although natural gas is still a fossil fuel with limited reserves, it would be a good transition fuel as we move toward cars that run on truly renewable energy.

The Honda Civic GX is the most popular natural gas passenger car currently on the market. You can refuel your natural gas car at stations in 45 states, or you can install a refueling hose in your garage (as long as you get natural gas delivered or pumped to your home, just as you would for your stove). Simply hook the hose up to your car for five to eight hours at night.

Use These Web Sites to Help Buy Your Next Car

- www.fueleconomy.gov—This is the Department of Energy's web site, created to help you find the car that has the best fuel efficiency and also fits your needs. Go to the "Find and Compare Cars" section to see fuel efficiency data and other information for any car made from 1985 to the present. Go to this section of the web site, www.fueleconomy.gov/feg/save money.shtml, to calculate how much your current car and the car you want to buy cost in gas each year and over their lifetimes.

- www.greenercars.com—This web site is home to the Green Book, a publication that rates all new cars based on their fuel efficiency (global warming pollution) *and* their tailpipe emissions (air pollution). You can find out the "greenest" and "meanest" vehicles in each class for free, but you'll need to pay $9 a month to access the full database.

Search For More Info

Hybrids

- www.howstuffworks.com/hybrid-car.htm—Go here to learn how gas-electric hybrid cars work.

Alternative Fuels

- www.eere.energy.gov/afdc—This is the Department of Energy's web site on alternative fuels. Go to this section of the web site, http://afdcmap2.nrel.gov/locator, to find out where alternative fueling stations are located across the United States.

Natural Gas

- www.ngvc.org—Check out the Natural Gas Vehicles Association's web site—it's loaded with great information.

Electric Vehicles

- www.electricdrive.org—Check out the Electric Drive Transportation Association's web site, full of information on electric vehicle legislation and new technology breakthroughs.

Fuel Cells

- www.howstuffworks.com/fuel-cell.htm—Go here to learn exactly how fuel cells work.

- www.fuelcells.org—This web site tells you about the latest technology breakthroughs in fuel cells. In fact, check out this entire web site for more great information on fuel cells.

Shop Right

Down Home Cookin'

The average meal travels 1,200 miles by truck, ship, and/or plane to reach your dining room table.

Overview. Here's a common scenario. It's February. The bananas you see in your supermarket were grown in Ecuador with the help of chemical pesticides and picked at the beginning of January so they could be yellow by the time they reached your store (the pesticides, by the way, were manufactured in a huge industrial facility that uses as much electricity as a full-size power plant). Once picked, the bananas traveled thousands of miles by plane, train, and truck, releasing thousands of tons of carbon dioxide into the atmosphere, until they finally arrived at your supermarket. Last, you drove your car to and from the store to buy them. Compare that with a local organic farm that doesn't use any pesticides and delivers produce to you every week in a reusable wooden box. Now, which option sounds better to you?

What You Should Know

- The Environmental Protection Agency reports that two-thirds of American farmers spray one billion pounds of pesticides onto our food crops each year.

- Organically grown food has been scientifically proven to contain more nutrients than the same food grown by conventional farming methods (i.e., using pesticides). Organic food contains, on average, 30 percent more magnesium, 27 percent more vitamin C, and 21 percent more iron than pesticide-grown food. More important, you're guaranteed that organic food has *100 percent* fewer pesticides and toxic chemicals than conventionally grown food does.

- After most pesticides are sprayed onto a crop, they remain active (which means they continue to release toxic vapors) for up to *one year*.

- There is twice as much carbon in the Earth's fertile soils as there is carbon dioxide in the atmosphere. The carbon in the soil is stored by millions of microorganisms that live in the soil and naturally keep it rich year after year. When chemical pesticides are used kill off "bad" pests, they also kill these vital microorganisms. As a result, their stored carbon gets released into the air, where it turns into the global warming gas carbon dioxide. Furthermore, after these organisms die, they can no longer keep the soil naturally fertile, so farmers are forced to use chemical fertilizers year after year in order to grow their crops.

- ***Climate Results:*** If, once a month, 100,000 people bought their weekly produce at a local farmers' market instead of at a grocery store, they would collectively prevent more than 3,000 *tons* of carbon dioxide from being emitted thanks to reduced transportation of the food.

- *Money Matters:* Although organic foods tend to cost slightly more than conventionally grown foods (about 20 cents more per pound), the more we buy them, the faster their popularity will grow and the faster their prices will become competitive. Also, spending a little extra money on organic foods now means you won't have to spend as much money later in taxes for agricultural pesticide cleanup or on medical bills for illnesses related to chemical ingestion.

Easy Ways You Can Help

- *Eat locally grown food.* When you buy locally grown produce or locally butchered meat, the benefits are endless. First, you help prevent global warming because your food doesn't travel across the country (or the continent!) in order to reach your kitchen table. Second, you support your local economy. Third, locally grown produce is fresher, better tasting, and more nutritious than transported produce (since nutritional value starts to decline as soon as food is picked or harvested). Locally grown produce is often cheaper and has less packaging. Last, by buying locally grown food, you promote your region's self-reliance and avoid supporting huge farming corporations that put their own profits over the environment. Shop at your area's farmers' markets and ask your grocery store manager to set up a special section for locally grown foods. If there isn't a farmers' market in your area, call up some local farmers and encourage them to start one!

- *Eat in season.* There's a reason why strawberries taste so good in June and cost a fortune in January. Although financially you may be able to buy fruits out of season, we're asking that you don't for the environment's sake. If you eat cantaloupe in January, it was most likely shipped by air, train, and truck from South America, Florida, or California. When you eat it in July, however, it was probably driven in from a neighboring county. Here's a basic guide of what to eat during each season (although this will vary by climate):

	Spring	Summer	Fall	Winter
Fruits	Blueberries	Bananas	Apples	Clementines
	Cantaloupe	Oranges	Asian pears	Grapefruit
	Cherries	Peaches	Coconuts	Grapes (Red)
	Pineapples	Plums	Cranberries	Kiwi fruit
	Raspberries	Tomatoes	Grapes	Passion fruit
	Strawberries	Watermelons	Tangerines	
Vegetables	Asparagus	Corn	Avocados	Chicory
	Carrots	Cucumbers	Beets	Kale
	Onions	Green beans	Broccoli	Radishes
	Peas	Peppers	Cauliflower	Snow peas
	Spinach	Summer squash	Leeks	Sweet potatoes
				Winter squash

- *Eat organically grown food.* Organic food is healthier for *you*, but we want you to eat it because it's healthier for the *planet*. Too many farmers today use chemical pesticides to grow crops bigger and faster instead of naturally controlling the pests. As a result, the billions of microorganisms that live in the soil and keep it full of nutrients die off. When they die, they release their stored carbon into the atmosphere, where it turns into carbon dioxide. Eventually, their bodies rot and release methane, a global warming gas that is 23 times more potent than carbon dioxide.

You can prevent these greenhouse gases from being emitted simply by supporting farmers and companies that grow their food organically, which means without the use of any chemical pesticides, fertilizers, or hormones, and in a sustainable manner (i.e., using crop rotation). A good way to start buying organic food is to switch to an organic variety of just *one* thing you eat frequently, such as carrots, coffee, or chicken. Thanks to its recent popularity, organic food is becoming more widely available and cost competitive every day.

- ***Plant a vegetable garden.*** If you want to completely eliminate your veggies' carbon dioxide emissions, why not plant a veggie patch in your backyard or in a plastic garden bed in the corner of or on the roof of your apartment? All you need is a little bit of water and sunlight. You'll be amazed at how little work a vegetable garden requires. See the third web site listed in "Search for More Info."

No Cost — *More Savings, Less Carbon Dioxide*

Annual amount of *money **saved*** as a result of buying 4 cantaloupes, 3 pints of strawberries, and 10 green peppers in season (June–August) as opposed to out of season (November–February): **$25**	Annual amount of CO_2 ***not emitted*** as a result of buying 4 cantaloupes, 3 pints of strawberries, and 10 green peppers in season (June–August) as opposed to out of season (November–February): **138 lbs.**

Assumptions: Cantaloupes cost $0.99 each in season, $2.99 each out; strawberries cost $0.98 pint in season, $4.00 out; green peppers cost $0.69 a pound in season, $1.99 a pound out; it takes 100 people doing this tip to prevent one truck full of produce that gets 5 miles per gallon from traveling 3,000 miles out of season; one pound of fuel burned = 23 pounds carbon dioxide emitted.

Search For More Info

- http://biodynamics.com/usda—Go here to find a local farm that delivers boxes of produce near you. Go here, www.ams.usda.gov/farmersmarkets/map.htm, to find a farmers' market near you.

- www.soilassociation.org/sa/saweb.nsf/resources/questions. html—Here are answers to your most frequently asked questions about organic food and farming practices.

- www.organicgardening.com/steps/new_garden.html—Learn how to plant your own organic garden here.

Eat Your Broccoli

The average American eats 260 pounds of beef per year, or 5 pounds of beef each week. One in five Americans are obese.

Overview. There's something you should know about cows and burping. When humans, cows, or any other animals burp or pass gas, methane is released. Up until now, we've been talking about the greenhouse gas carbon dioxide. Methane is another greenhouse gas that is more rare than carbon dioxide, but 23 times more potent in terms of global warming. In other words, even though there's much less methane in the atmosphere than there is carbon dioxide, it can trap 23 times more heat inside the Earth's atmosphere than carbon dioxide can.

Now, because humans and most other animals don't burp or pass gas that much, this methane emission is not a serious problem. But because cows have four stomachs and a complicated digestive system, they burp *an awful lot*. In one year, the world's 1.3 billion cows burp up 75 *million tons* of methane! In addition to all this burping and passing gas, the colossal number of cows we raise for beef production creates huge piles of manure—which also gives off methane. Last, it takes *enormous* amounts of energy, water, and grain to raise and then slaughter beef cattle. In fact, in the United States alone cattle eat enough grain and soybeans each year to feed one billion people for a year. Now that you know all this, is eating one fewer Big Mac a week really too much to ask?

What You Should Know

- Livestock in the United States currently consume 70 percent of all the wheat, corn, and grain we grow and graze on pastures that take up 35 percent of our total land area.

- In addition to the methane that's burped up by cattle, carbon dioxide is released when forests are clear-cut to make room for grazing land. When these trees are killed, the carbon they have been soaking up and storing for decades is released into the air, where it turns into carbon dioxide. Twenty-five million acres of rainforest have already been cut down in Brazil alone to make room cattle pastures, and the number keeps growing.

- Your typical Quarter Pounder with cheese represents 55 square feet of rainforest that has been clear-cut to raise cattle.

- For every one pound of beef produced, seven pounds of grain must be fed to the cow and 2,500 gallons of water must be used to grow that grain. If Americans reduced their meat consumption by only 10 percent, they would save enough grain to feed 60 million people—which is the number of people who starve to death each year.

- The electricity needed to water, feed, slaughter, and package just one pound of beef is the equivalent of burning one gallon of gasoline in your car's engine.

- Each year, the 260 pounds of steaks and hamburgers that *you alone* eat are responsible for nearly 1.5 *tons* of carbon dioxide (or 130 pounds of methane) being emitted during the life of the cattle.

- Of all the methane emitted each year as a result of human activities, 24 percent comes from our cattle's belches and manure. Methane is also released by nature, for example, in bogs and marshes, but because of nature's balanced cycle, the amount of methane emitted naturally each year is soaked up by other plants and organisms.

- Beef cattle are not the only methane belchers—other grazing animals such as sheep and pigs burp up methane too, but not nearly as much as cows do. No studies have been done yet on poultry or fish, but it's safe to assume that they emit less methane than cows, since they don't have four stomachs like cows do. The moral of the story? Eating chicken and fish is better for global warming prevention than eating pork and lamb, which is better than eating beef.

- Termites are the only animals that produce *more* methane than cows do, and they're the only animals that produce *as much* methane as the entire human industrial sector does. Because of this, they are a serious contributor to global warming! However, since humans already do everything they can to prevent termites (and don't foster them they way we do cattle), we can only hope that science will find a better way to control termites. Choosing not to eat a hamburger this week, though, is something you *can* do.

- *Climate Results:* If 100,000 people reduce the amount of beef they eat each year by 25 percent, they'll prevent 1,800 *tons* of carbon dioxide from being emitted each year. In addition, they'll save seven million pounds of grain and four billion gallons of water each year.

- *Money Matters:* Growing fruits, vegetables, and grains requires 95 percent fewer raw materials than raising and slaughtering cattle does. In fact, the total value of raw materials used for meat production in the United States in one year is higher than the value of all the gas, oil, and coal consumed by the United States in one year.

Easy Ways You Can Help

- *Eat less meat.* Cutting back on the amount of meat you eat doesn't mean you have to become a vegetarian (although there are currently 7.5 million vegetarians in the United States). Start off slowly—try to cut down to five meals with meat each week. After six months, cut down to four meat meals a week. After a year, see if you can eat meat only two times a week, preferably eating fish or chicken instead of livestock. On a high-protein diet? By all means, stick with it! Simply get your protein from the dozens of protein sources that aren't beef, such as beans, nuts, and tofu. Besides helping prevent global warming and world hunger, you'll be helping your body: Eating less meat has been proven to reduce the risk of heart disease, breast cancer, prostate cancer, ovarian cancer, and colon cancer.

- **Know how you're helping.** Right about now, you're probably thinking something like this: If I go out to a restaurant and decide not to order a steak, someone else is just going to order one and the steak will be eaten anyway! *Not necessarily.* Restaurants and grocery stores run on a supply-and-demand principle. Perhaps other people are, like yourself, beginning to realize the negative environmental impacts of eating beef. Maybe no one will order the steak tonight, or the next night, or the next night.

  The chef will be forced to bring home the steak to his family, and his wife will start to complain, "This is the third night in a row we've had steak! What's going on, no one likes steak anymore?" As a result, the next week when the chef places his order with the butcher, he will order less steak. The same thing could happen at your grocery store. So know that when you decide not to order a steak, you *are* making a difference.

- **Solve world hunger.** Right about now, you're probably *also* thinking something like this: Sure, if I eat less meat, there will theoretically be more grain available to feed the hungry, but I know that won't really happen. So why bother if it won't really make a difference? That's a great question! We're going to need to see a major change in agriculture if we want to really do something about world hunger. Here's one possibility: By eating less meat, you'll save money. You and every other American who eats less meat could then donate some of that saved money to a fund, run by the government, that pays farmers to keep growing their wheat. But, instead of feeding that wheat to cattle, they could bundle it up and send it to those who are hungry.

Write to your congressional representative today (see Tip 51) and tell him or her why you've started to eat less meat. Ask him or her to figure out a way to use the resulting grain saved to feed the hungry. Ask the government to stop paying farmers *not* to grow crops (called subsidizing), and instead pay them that money to grow crops and send the food to hunger-plagued regions.

- *Take cooking classes.* The hardest thing about gradually cutting meat out of your diet is learning what to replace it with. Splurge on a good vegetarian cookbook, such as Mollie Katzen's classic *Moosewood Cookbook* and its sequels to learn how you can make a complete meal that doesn't center around meat. Better yet, ask to be enrolled in vegetarian cooking classes for your birthday.

- *Wait for science.* Scientists are currently working on ways to change the diets of cattle so they don't burp up as much methane. They're even trying to find a way to breed methane-free cattle! Our advice is to cut down on your meat intake *now*, and when science pulls through, *then* order that triple cheeseburger.

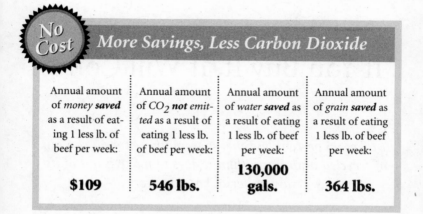

Assumptions: *Average American eats 260 pounds of beef per year; eating 65 percent ground beef and 35 percent steaks averages to six dollars per pound of beef; the food that replaces the beef costs 35 percent less on average; 1 pound beef = 10.5 pounds carbon dioxide equivalent.*

Search For More Info

- www.epa.gov/globalwarming/greenhouse/greenhouse15/cow.html—At this web site, the Environmental Protection Agency describes one way science is trying to deal with cows and all their methane.

- www.pleasebekind.com/veg/faq.htm—Here are answers to your most frequently asked questions about what it's like to be a vegetarian.

If You Buy It, It Will Come

It takes 95 percent less energy to manufacture a can out of recycled aluminum than it does to make it out of its raw material bauxite ore.

Overview. The United States may not be a very environmentally conscious country, but if there's one thing we're good at, it's capitalism. In fact, our economy is so sophisticated that major retail chains now have computers that monitor exactly how many products are sold each day and from which stores. In other words, every day, with every purchase you make, someone is watching. And when you ask the manager of your grocery store to stock a certain item because it's made of recycled material, he'll tell that to his suppliers, who will then tell their superiors, and suddenly you've started a trend. That's how capitalism works: demand dictates supply. Manufacturers in the United States aren't making environmentally friendly products for one reason and one reason only: We're not asking for them.

What You Should Know

- You save energy when you buy a recycled product over a product made of virgin materials because it *always* takes less energy to manufacture a recycled product than it does to make a brand-new one (with the exception of a few mixed-plastic products). You also directly support the recycling industry and all the people it employs, conserve natural resources, save landfill space, and encourage those manufacturers who don't make the environment one of their top priorities by selling recyclable goods to make it one—and fast!

- Not every product that's made of recycled content has a label to tell you that. Until manufacturers start realizing that consumers *want* to know what their products are made out of, keep in mind that these four types of packaging *always* contain a fair percentage of recycled content: aluminum soda cans (at least 50 percent recycled content), glass containers (at least 25 percent), tin cans (actually tin-coated steel, at least 25 percent), and pulp cardboard (e.g., egg cartons, cardboard flowerpots, 100 percent).

- The climate-neutral craze (see the following section) has already hit Europe—in England you can buy a carbon-neutral car or vacation package, or take out a climate-neutral mortgage! When Mazda released its new car, the Demio, in the United Kingdom, this environmentally conscious company announced that they would plant five trees for every car sold in order to make up for the carbon dioxide that the car would emit during its first year. After that year, Mazda encouraged its customers to continue planting five trees a year.

- ***Climate Results:*** If you're planning on recarpeting your home this year, you'll prevent one *ton* of carbon dioxide from being emitted if you recycle your old carpet and replace it with partially recycled new carpet from Interface (based on an average of 1,600 square feet of carpeted space per home).

- *Money Matters:* Recycled paper (this includes spiral notebooks and Post-it notes as well as printer and copy paper) is equal in quality to but often 10 to 30 percent cheaper than nonrecycled paper.

Easy Ways You Can Help

- *Buy climate-neutral!* Over the next couple of years, you'll start to see more and more products labeled "climate-neutral" or "climate-cool," just like when we saw cans of hair spray labeled "CFC-free" in the 1980s. Climate-neutral means that a product was delivered to your store or home with no *net* greenhouse gas emissions. Here's how it works: A company that wants to label its product climate-neutral must first do everything it can to reduce the carbon dioxide emissions that occur during manufacturing and transportation of the product. This could mean anything from cleaning up the factory's smokestacks to making the executive offices use recycled paper.

  Once the Climate Neutral Network's advisory panel has determined that a company has done everything in its power to reduce emissions internally, the company is allowed to cancel out its remaining carbon dioxide emissions by investing in outside projects that will result in reduced emissions (with the stipulation that 66 percent of these projects must occur within the United States). For example, two companies,

Shaklee and Interface, canceled out their remaining carbon dioxide emissions by providing funding to a local public school that needed to replace its 65-year-old heating system.

- *Support climate-neutral companies.* The environment benefits when a company decides to go climate-neutral because less carbon dioxide will be emitted. But the company also wins, since that climate-cool label earns the loyalty of thousands of environmentally conscious consumers. You can encourage all businesses to go climate-neutral by supporting those that have already earned the label.

  Currently there are only seven companies that have met the Climate Neutral Network's strict climate-cool criteria, including Shaklee (health products), TripleE (air travel/travel agency), and Interface (carpeting) (see www.climateneutral.com for the complete list). The Climate Neutral Network is also working with the following companies to develop climate-neutral gasoline, clothing, vehicles, and more:

  - Delta Air Lines
  - Ben & Jerry's
  - Nike
  - Sunoco
  - BP Amoco
  - Chevron
  - Mead Corp.
  - Cognis, U.S.
  - The Body Shop
  - Toyota
  - Philips Electronics

Four companies—Earthbound Farm (organic food), Stonyfield Farm (yogurt), Shaw's (grocery stores), and Kinko's (photocopying)—also invest in carbon-dioxide-reducing projects to offset their emissions, although they haven't yet applied to be certified by the Climate Neutral Network. Also, Patagonia (outdoor clothing) gives grants to nonprofits that help prevent global warming. You can do your part by supporting all these climate-conscious companies.

- ***Buy recyclable.*** Now that you've looked for the climate-neutral label, here's another thing you should look for before buying something—check to see if the product and/or its packaging can be recycled. Look for the recycling logo (right), and always shop with the following order of preference in mind: glass is easiest to recycle, next is aluminum, then paper, then plastic; products made of multiple materials are the hardest to recycle. Last, plastic containers with the number 1 or 2 in the center of their recycling logos are the easiest to recycle, as opposed to ones with 3, 4, 5, or 6 in the logo.

- ***Buy recycled.*** The recycling business is thriving, and as a result we have developed ways to make products from recycled materials that are equal in quality to and often less expensive than products made from raw materials. Always compare the labels of similar products to see if one has a higher percentage of recycled content (also called postconsumer content). Here are just some of the recycled products you can buy today:

Paper
Newspapers
Printer paper
Paper towels
Paper napkins
Toilet paper
Facial tissues
Egg cartons
Cereal boxes
Shoe boxes
Cardboard boxes
Manila folders
Envelopes
Greeting cards
Comic books

Plastic
All plastic bottles
Plastic toys
Plastic garbage bags
Plastic garbage cans
Recycling bins
Stuffing for pillows
Sleeping bags
Synthetic fleece
    clothing
Garden hoses
Floppy computer disks
Car bumpers
Carpet (partially made
    of recycled plastic)

Glass, Steel, and so on
Glass bottles
Glass containers
Anything glass
Aluminum cans
Nails
Steel or tin-coated cans
Anything steel
Paint
Re-refined motor oil
Wall insulation
Pressed wood
Roofing tiles
Car tires (partially made of
    recycled rubber)
Sawdust pellets (for your
    fireplace)

No Cost
More Savings, Less Carbon Dioxide

Annual amount of *money* **saved** as a result of buying two gals. of reblended paint (same quality, but made with 80 percent recovered paint) instead of new paint:

$15

Annual amount of CO_2 **not emitted** as a result of buying 2 gals. of reblended paint (same quality, but made with 80 percent recovered paint) instead of new paint:

37 lbs.

Assumptions: Reblended paint costs 44 percent less than standard paint; standard paint costs $17 per gallon; 1 gallon of paint weighs 200 ounces; creating 1 pound of consumer product creates 5 pounds of waste; 1 pound of waste = 1 pound of carbon dioxide emitted; manufacturing reblended paint uses 30 percent less energy than manufacturing new paint.

Search For More Info

- www.climateneutral.com—Go to the Climate Neutral Network's web site to learn more about what it means to be climate-cool and to see which companies have been recently certified.

- www.responsibleshopper.org—This fantastic web site lets you find out how environmentally responsible certain companies and industries are.

- www.gaiam.com—Go here and search for "recyled" to buy a variety of recycled products online.

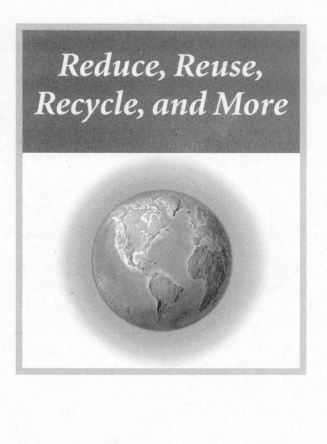

Reduce, Reuse, Recycle, and More

Less Is More

The typical American creates 4.6 pounds of garbage per day, or 1,700 pounds of garbage per year—that's double what it was thirty years ago.

Overview. Did you ever wonder why it's "Reduce, Reuse, Recycle" instead of "Reuse, Recycle, Reduce"? Well, it's because they're in order of importance. *Reducing* means buying less stuff and therefore throwing away less stuff. It's the most important of the three R's because it stops waste from even happening—if you don't buy as much, you don't need to reuse, recycle, or throw away as much. But what does reducing have to do with global warming? you might ask. A lot. Energy is needed to collect, process, manufacture, and ship *everything* you buy. In today's world, that means fossil fuels burning and emitting carbon dioxide. In the future, we hope we'll use solar and wind energy to power our factories, but until that time comes, keep in mind that everything you own was created by the burning of fossil fuels. So, what's the single best way you can help reduce those carbon dioxide emissions? Reduce the amount of stuff you buy.

What You Should Know

- We just told you how reducing the number of things you buy and throw away will conserve energy and prevent carbon dioxide emissions, but what's more obvious is that reducing will save you money—when you buy less, you spend less!

- Packaging materials make up one-third of America's landfills.

- The typical American receives 550 pieces of junk mail a year.

- Seven out of 10 women rank shopping as the number-one way to cheer themselves up. Imagine how much carbon dioxide emission would be prevented if they called a close friend instead.

- If everyone in the world lived the way the average American does, we would need four more planet Earths to provide all the necessary materials and energy.

- ***Climate Results:*** If you register with the web site listed in the following section and therefore decrease your junk mail by 75 percent, you will prevent 171 *tons* of carbon dioxide from being emitted each year.

- ***Money Matters:*** Assuming the average three-person household spends $75 a week on groceries, skipping a grocery trip once every three months in order to use up all of the food in the kitchen will save that household around $300 a year.

Easy Ways You Can Help

- ***Become a minimalist.*** Americans get to choose from 200 types of breakfast cereals and over 5,000 styles of shoes, while many people in this world can't even afford a loaf of bread. It's time for us to end our love affair with consumerism before we do permanent damage to our planet. Go through your house today and donate or recycle everything you don't really need (see Tips 45 and 47)—then leave it that way! Stores respond to the demands of their customers: The less we fill our closets and cupboards with things, the less our factories will burn fossil fuels to manufacture those things. Besides, when you look back on your life, which will you remember more—the shoes you had or the friends you made?

- ***Buy in bulk.*** Buy large quantities of things—especially non-perishable goods such as shampoo, toilet paper, or peanut butter—whenever possible. Bulk items use less packaging, which means less energy is needed to manufacture them. Buying in bulk is also much cheaper—about 50 percent, on average! For example, buying six small juice boxes costs twice as much as making a pitcher of juice from concentrate. And individually wrapped servings of oatmeal cost more than *three times* as much as the same amount of oatmeal from a large container. To make things less cumbersome, save the smaller shampoo bottle (or any other container) you have now, buy a bulk size to store in the cupboard, and keep refilling your smaller bottle.

- *Say no to excess packaging!* Energy is needed to produce the aluminum wrapper that surrounds your stick of gum, *and* the paper sleeve that surrounds the aluminum wrapper, *and* the paper wrapper that holds the sticks of gum together. When you choose to buy products with the least amount of packaging, companies that squander energy with wasteful wrappers will start to see a drop in sales. Eventually they'll conduct surveys to determine why their sales have dropped, and when they find out it's because of their excess packaging, they may well change their ways.

  Watch out for companies, such as makeup manufacturers, that purposefully add packaging to catch your eye. Also, stay away from individually wrapped foods, like TV dinners or Kraft Singles cheese; instead, go for something healthier, such as fruit that has no packaging at all. When faced with two equal products, always buy the one with the least packaging— besides helping the environment, it will probably cost less!

- *Stop junk mail once and for all.* Did you know that the U.S. government *allows* junk mail to be sent to you? Just think of all the trees that are cut down—releasing their carbon dioxide back into the atmosphere—and then all the energy that's used to manufacture America's *4.5 million tons* of junk mail each year! Simply fill out the on-line form at this web site to get off of most mass-mailing lists for five years (go to www.the-dma.org/cgi/offmailinglistdave. There's a five-dollar on-line processing fee, or you can mail in a form for free). Your junk mail will decrease by up to 75 percent! Also, see if you can have your bills e-mailed to you so you can pay them on-line. You'll save paper and stamp money.

- ***Don't be cheap.*** Save up so you can buy quality goods instead of cheaper products that will wear out and need to be replaced sooner. You'll end up saving money (and energy) in the long run. This is especially true for furniture and major appliances.

- ***Empty out the fridge.*** Every few months, skip your weekly grocery trip and spend the week cooking what's buried in your fridge and cupboards. If you find something good, you just saved money and energy. If you find something that's gone bad, isn't it better that you found it sooner rather than later?

- ***Use less in general.*** There are a million ways to use less stuff: get a library card instead of buying books (attention, college students), buy fewer toys and do more arts and crafts projects with your kids, or simply make a vow to use a bag at a store only if you can't carry the item or put it in your purse. You don't need to be a creative genius—just step back and look at how much stuff you use and buy. Then ask yourself, Do I really need that much?

No Cost

More Savings, Less Carbon Dioxide

Annual amount of *money **saved*** as a result of buying 15 percent of your groceries in bulk sizes instead of smaller sizes:	Annual amount of CO_2 ***not** emitted* as a result of buying 15 percent of your groceries in bulk sizes instead of smaller sizes:
$293	**1,354 lbs.**

Assumptions: Buying in bulk costs on average 50 percent less than buying individual or smaller packaging; a typical family of three spends $75 a week on groceries; groceries and their packaging account for 30 percent of a family's yearly waste, each person creates 4.6 pounds of waste a day; less packaging reduces the waste of a product by one-third; one individually wrapped grocery item needs eight kilowatt-hours to be manufactured, with less packaging it needs only four kilowatt-hours.

Search For More Info

- www.costco.com—Shop online at one the nation's largest retailer of bulk goods.

- www.the-dma.org/consumers/offmailinglist.html—Here are answers to your most frequently asked questions about why junk mail happens and how to stop it. Also, go here for more great tips on how to stop junk mail: www.coopamerica.org/woodwise/consumerguidejunkmail.htm.

- www.epa.gov/epaoswer/non-hw/muncpl/sourcred.htm— Check out the Environmental Protection Agency's web page on reducing your waste (they call it source reduction).

Use It Again, Sam

Every year, Americans throw away 25 billion plastic foam cups.

Overview. So you've cleaned out your closets and stopped buying individual juice boxes. Now that you've reduced as much as you can, it's time for the second most important R—Reuse! *Reusing* means using a product more than once, whether for the same purpose or to serve a new purpose. It's better than recycling because the product doesn't need to be remanufactured before it can be used again. Tremendous amounts of carbon dioxide are prevented from being emitted when a product doesn't have to be disassembled, shredded, melted, and re-formed the way it does when it's recycled, or when it doesn't need to be transported to a landfill in an emissions-spewing garbage truck. Reusing also means you don't have to buy a replacement product—which needed a lot of energy to be manufactured. Best of all, reusing things saves you surprising amounts of time and money!

What You Should Know

- From 1950 to 2000, Americans consumed as much food and material goods as all the generations of Americans before them.

- More than 70 percent of all drinks in Germany are sold in refillable glass bottles—not because they don't have the technology to make plastic bottles but because their government has decided to do what's best for the environment. In the United States, only 5 percent of all beer and soft drinks are sold in refillable glass bottles.

- A typical household uses about 700 paper grocery bags in two years—that's equivalent to all the wood in a 20-year-old tree. Meanwhile, one canvas grocery bag lasts approximately five years.

- Americans throw away 570 disposable diapers every *second*.

- Since 1973 students at the University of Arizona have been getting academic credit for digging through our landfills. Besides learning what Americans throw away, they've made another surprising discovery—our landfills are so tightly packed that there's no room for water or air to do the job of decomposing. As a result, they've found 15-year-old undecomposed hot dogs. Imagine how long it takes our so-called biodegradable plastic bags to decompose!

- ***Climate Results:*** Each day Americans drink 45 million cups of coffee or tea in disposable cups. If every coffee-drinking American switched to using a reusable thermos, they would prevent seven *million pounds* of carbon dioxide from being emitted *every day*, thanks to the energy saved from not having to manufacture and recycle or dispose of in a landfill all those paper cups.

- *Money Matters:* If your baby is in diapers until he or she is 1½ years old, using washable cloth diapers instead of disposable diapers will save you approximately $900, even after taking into account the cost of the cloth diapers, bed covers, liners, detergent, and hot water (for the washing machine).

Easy Ways You Can Help

- *Use cloth bags.* Paper or plastic? Neither! Do what the Europeans have been doing for years—invest in some durable cloth bags and reuse them. Keep them in your trunk and then throw them in the bottom of your cart while you shop. Also, save those little plastic bags from your fruits and vegetables and reuse them on your next shopping trip. Since you should always wash your produce anyway, reusing the plastic bags will be just as sanitary as using new ones.

- *Use a thermos.* Spend a couple of bucks on a thermos and tell Starbucks to fill 'er up—they'll be happy to help the environment! Buy a liter of Coke and fill up your thermos every day instead of going through four aluminum cans. Make a pitcher of juice and fill up your thermos instead of throwing away six juice boxes.

- *Buy reusable goods.* Buy rechargeable batteries instead of disposable ones, a $20 reusable camera instead of a disposable one, and an ink refilling kit instead of a disposable printer cartridge. Some other examples of reusable goods are reusable metal coffee filters, razors with refillable blades, and refillable pens and pencils. In general, never buy anything that says "disposable."

- ***Reuse those floppies.*** Computer users around the world throw away four million floppy disks *a day*. That's because they don't realize that floppy computer disks (as well as videotapes, and audiotapes) can be written or taped over *thousands* of times. In fact, most floppy disks today come with a lifetime warranty. To erase a floppy so that you can reuse it, right-click your computer's A drive icon, select "Format," and then select "Full" format.

- ***Get back into bottles.*** Sometimes it *is* a good idea to be old-fashioned—and glass soda bottles are a perfect example. Since the 1950s, Americans have moved away from refillable, glass bottles to the throwaway, plastic ones. But those plastic bottles don't just disappear—they sit in our landfills! Ask your grocery store manager to start stocking glass bottles and to create a return depot.

- ***Break out the good china.*** Instead of paper plates, always try to use ceramic plates that can be washed with water (which is renewable). Paper or plastic plates may be easier on *you*, but they're much harder on our *landfills*. When you must use plastic plates or silverware, buy the thickest ones you can find, then wash and reuse them—some are even dishwasher safe! For an environmentally friendly lunch, use a lunch box instead of a paper bag, a cloth napkin instead of a paper one, a thermos, and reusable plastic food containers and silverware.

- ***Get crazy about cloth.*** Eighteen *billion* disposable diapers are dumped into U.S. landfills each year, and each one takes up to 500 years to decompose. You may not be around to see the true effects of global warming, but your children will be.

For their sake, switch to reusable, washable cloth diapers. Also, always use cloth napkins, handkerchiefs, and rags instead of paper napkins, facial tissues, and paper towels.

- *Give "green" gifts.* Save and reuse as many boxes, bows, ribbons, Styrofoam packing peanuts, and tissue and wrapping paper as you can (unwrap slowly so you don't wrinkle it!). If you have more packing peanuts than you need, call the National Peanut Hotline (800-828-2214) to find the location nearest you that accepts peanuts for free (e.g., Mail Boxes Etc. recycles these). When you're the one buying the gift, keep in mind that colorful gift bags are easier for your friends to reuse than boxes with wrapping paper are. Also, try tying your boxes with string instead of using tape. It's easier to reuse and cheaper. Last, be creative: Use fabric or your kids' artwork instead of wrapping paper.

- *Become Mr. or Ms. Fix-It.* When something breaks, don't immediately throw it away. First, see if you can fix it yourself—for example, you can easily repair most scratches on CD's with a rag and some toothpaste. If that doesn't work, you can buy an inexpensive refinishing kit. If you can't fix something yourself, decide whether it makes more sense to pay a professional to fix it than to buy a new one. Look in the yellow pages under the name of the item and "Repairs."

- *Reuse, reuse, reuse!* Here are just a few ways you can reuse things around your house:
  - Write messages and lists on the back of scrap paper
  - Rinse and reuse zipper-lock bags and aluminum foil (except if they were used for meat or dairy products)

- Reuse envelopes by sticking labels over the old addresses
- Tie together your mesh onion and orange bags to use as scouring pads for your dishes
- Use leftover wallpaper scraps to wrap presents, make home-made greeting cards, or line cabinets and drawers
- Use your excess coffee and tea to water your plants—the minerals are good for them!
- Cut up old clothing and sheets to use as dustrags
- Rinse out and reuse glass jars, milk jugs, and plastic containers (such as margarine or yogurt tubs) to store food, flour, buttons, and so on
- Reuse plastic and paper grocery bags as trash can liners
- Wrap packages you want shipped with paper grocery bags
- Cut the tops off large detergent boxes and use the empty boxes to store file folders
- Cover an empty coffee can with pretty paper, add dirt, and make it into a flowerpot
- Reupholster your old couch instead of getting a new one
- Place old panty hose in the bottoms of flowerpots for better drainage

Low Cost

More Savings, Less Carbon Dioxide

Annual amount of *money* **saved** as a result of switching from paper towers and napkins to rags and cloth napkins, taking into account the cost of the cloth napkins:

$45

Annual amount of CO_2 *not emitted* as a result of switching from paper towels and napkins to rags and cloth napkins:

214 lbs.

Assumptions: Based on typical household usage of two paper towel rolls per month and six paper napkins per day; paper towels cost $6.89 for six rolls, paper napkins cost $3.19 for 200; six cloth napkins cost $3.00 each; rags are free (cut up old clothing); cost of laundering napkins is no more than normal weekly loads; manufacturing emits carbon dioxide equal to five times the weight of the product; waste disposal emits carbon dioxide equal to the weight of the product.

Search For More Info

- www.enn.com/enn-features-archive/1999/05/051399/bags_3170.asp—Check out this man's opinion of which cloth grocery bag is the best. Go here (www.ecobags.com/product-spg.asp), to buy cloth grocery bags on-line.

- www.diaperpin.com/howto.asp—This fantastic web site explains how to make the transition from disposable to cloth diapers. Important note: If you live in a region where water is scarce, it may be better for the environment for you to use disposable diapers so you do fewer loads of laundry.

- www.redo.org—This is a great web site about reusing things and what's happening in the United States to promote reusing.

Be Thrifty

Americans hold approximately 55,000 yard sales a week and spend two billion dollars a year shopping at them.

Overview. Okay, so you've reused the yogurt container to keep your leftovers, but you really, *really* want to chuck Aunt Susie's hideous lamp. Well, you're in luck. You can get rid of it without throwing it away. That's because somewhere, someone is dying to have a lamp just like Aunt Susie's. So, you look in the yellow pages and find your nearest Goodwill center or thrift shop, or you sell it on eBay. Even though *you're* not reusing the lamp, someone else is, which means you don't have to throw it away to be sent to a land-fill or recycle it—both of which use energy. That's the beauty of reusing—you can give or sell things you no longer want or need to those who can still get some use out of them. As an added bonus, you'll prevent more carbon dioxide from being emitted!

What You Should Know

- There are more than 1.8 million on-line auctions for used goods held *every day* on eBay (www.eBay.com).

- Selling your used things *earns* money for you and *saves* money for the people who are buying them (since they don't have to pay full market price). By contrast, throwing your things away will cost money for disposal *and* hurt the environment.

- The waste created by manufacturing one laptop computer weighs 4,000 times the weight of the finished laptop. If you're buying a new computer, be sure to donate or sell your old one.

- There are over 6,000 reuse centers in the United States, including 1,800 Goodwill donation drop-off centers and even more Salvation Army drop-off boxes.

- Clothing and other items sold at thrift or secondhand stores cost on average 35 percent less than brand-new items.

- It's not uncommon for a well-planned, two-day garage sale to earn $500 to $1,000.

- In a little town called Fontana, Wisconsin (where one author happens to be from), two things were happening: the local church needed to raise money for a new organ and too many people were throwing out their old clothes. As a result, four retired ladies decided to turn the attic of the church into a thrift shop. Two years later, the new organ sounds beautiful, some women who were bored now have full-time jobs where they can chat all day long, and hard-pressed families can buy shirts, dresses, pants, shoes, purses, sweaters, and jackets for two dollars each.

- ***Climate Results:*** If 100,000 people donated their old couches to charity instead of putting them on the curb to be land-filled, they would collectively prevent 30,000 *tons* of carbon dioxide from being emitted (taking into account all the energy used to manufacture new couches and to landfill old couches).

- ***Money Matters:*** Donating things can lower your income taxes! That's because when you make a donation (either in cash or in goods) to a recognized nonprofit organization, such as the Salvation Army, Goodwill, or your local church or school, you get to deduct the dollar value of your donation on your annual income taxes.

    Here's how it works: If your household's taxable income this year is $50,000, then your federal tax bill will be 28 percent of that amount—$14,000. However, if you donate $2,000 worth of stuff to Goodwill this year, the IRS will subtract that from your gross income, bringing the taxable amount down to $48,000. Even though you're still in the 28 percent tax bracket, your tax bill will now be only $13,440. That's a tax savings of $560—just for donating stuff you didn't want anymore! Three things to keep in mind: You must get a receipt for your dona-tion, your calculations must reflect the *current* market value of the goods (not the price you originally paid for them), and you must itemize the deduction on your tax form.

Easy Ways You Can Help

- ***Take it back.*** Ask your supermarket if you can bring back your empty egg carton each week to be reused by the egg sup-pliers. Take any extra wire hangers back to your dry cleaners to be reused instead of throwing them away. Return produce baskets to farmers' markets for reuse. If you live in a state with a bottle deposit rule (go here to find out: www.epa.gov/epaoswer/non-hw/muncpl/mswdata.htm#item1), bring your empty bottles back to the store where you bought them to receive 5 to 10 cents per bottle.

- *Have a yard sale!* This is a great way to pass the stuff you no longer need on to people who can use it, and make a little money in the process. Clean out your basement, make some homemade price tags, and hang up some flyers or put an ad in the newspaper. You'll be surprised at how many people show up and how much money you can make. You could also sell your things to an antiques shop or trade them in at a used goods store for other products.

- *Discover the wonder of eBay.* Selling and buying used goods at on-line auction web sites has become insanely popular—and for good reason. They're like giant national yard sales! But you won't find just old lamps at these Internet sites—you'll also find laptop computers, VCR's, snowmobiles, and more. Two of the most popular auction sites are www.ebay.com and www.ubid.com.

- *Be a thrifty shopper.* Don't just *have* a yard sale, shop at one! Look for announcements of upcoming sales in your local classified ads. Also, be sure to check out local thrift stores, consignment shops, used bookstores, and antiques shops before heading to the mall—besides saving money, you'll be saving all the energy that would have been used to manufacture brand-new items.

- *Donate!* If you don't have time to plan a yard sale, don't just throw your things away. There are plenty of organizations and individuals that could benefit greatly from your belongings. Some of these groups will even pick up items from your home. You'll be helping the environment by delaying the items' entrance into our waste stream. Here are just a few of the many things you can donate to your community:

Donate Things Like

Clothes	Computers, printers, fax machines
Food	Electronic equipment, audiotapes, videos
Books, magazines, records, CD's	
Appliances	Lumber, tools
Furniture, carpeting	Arts and crafts supplies (fabric, egg cartons, aluminum pie plates, etc.)
Office supplies and scrap paper	

To Your Local

Salvation Army or Goodwill drop-off center	School, public library
Homeless shelter, food bank, soup kitchen	Hospital, doctor's office
	Nursing home
Charity, orphanage	Nonprofit business
Low-Income housing development	Thrift store
Church	Used goods store

- *Donate your vehicle!* Often it makes more sense for you to donate an old car or other vehicle than to try to sell it. That's because a nonprofit organization will give you a receipt for the absolute highest current value of the car, which you can then deduct from your income taxes. You'll have a harder time selling the car for that amount of money. Call 1-800-488-CARS or go here, www.kidney.org/funds/cars/cfm, to donate your car, boat, truck, or van—*running or not*—to the National Kidney Foundation. It's fast, tax-deductible, and they'll pick up your vehicle for free.

- *Start a thrift shop.* If your town doesn't have a thrift shop or donation depot, start one. If you don't have the time, encourage a budding entrepreneur to either open a franchise of Salvation Army or Goodwill or start his or her own business. Think of other ways to reuse on a community level—organize a program through which residents and businesses can donate their used materials to schools for arts and crafts or for an annual school fund-raiser. Organize a toy or sports equipment exchange for older kids who grow out of things and younger kids who could use them. Enlist your friends and neighbors to help!

Low Cost

More Savings, Less Carbon Dioxide

Average amount of *money **earned*** as a result of having a one-day yard sale to sell your unwanted belongings, taking into account the cost of advertising:

$150

Average amount of CO_2 **not** *emitted* as a result of having a one-day yard sale to sell your unwanted belongings.

1,500 lbs.

Assumptions: Based on $300 average profit for a two-day yard sale after spending $25 on advertising; assume 250 pounds of items are sold in one day; for every pound of waste eliminated from the waste stream, carbon dioxide emissions are reduced by one pound; manufacturing one pound of brand-new goods creates five pounds of waste.

Search For More Info

- www.yardsalequeen.com—This wonderful web site gives tips on how to throw a successful yard sale, advice on how to shop at yard sales, and much more.

- http://locator.goodwill.org—Type in your zip code to find the Goodwill Donation drop-off center nearest you.

- www.satruck.com/FindDropoff.asp—Type in your zip code to find the Salvation Army drop-off center nearest you.

Compost Most Everything

Over two-thirds of the total waste produced in the United States is compostable, yet we still send 30 million tons of leaves and grass clippings to landfills each year.

Overview. The 1990s was the decade of recycling—environmental groups created awareness, recycling rates started to rise, and soon a full-fledged industry was born. We're hoping that the first decade of the new millennium will be the decade of composting. Composting occurs when microorganisms break down (decompose) organic material such as grass, leaves, or food and turn it into rich soil. In other words, composting is nature's way of recycling.

What does composting have to do with global warming and saving money? Well, the best thing about composting is that it keeps organic material out of landfills. When organic material is crushed deep in a landfill, it releases methane, the most potent greenhouse gas (23 times more potent than carbon dioxide!). Composting can also save money. If you currently pay for garbage disposal according to the amount of trash you throw out, composting can help you reduce your annual trash bill by *at least* 25 percent. Plus, thanks to the rich soil that composting produces, you won't have to buy as much fertilizer or mulch for your lawn or garden. Best of all, composting is easy!

What You Should Know

- Composting further lowers greenhouse gas emissions because it reduces the amount of garbage that needs to be transported (by gasoline-powered trucks) to a landfill. Remember how we said that a lighter car uses less gas? Well, a garbage truck will be lighter thanks to your composting! Also, if you spread the fertile soil that composting produces—called humus—onto your garden or lawn, it will work just like trees do to soak up and store carbon dioxide that's in the air (see Tip 31).

- The key to having a compost pile that doesn't give off an odor is oxygen. There are two kinds of microorganisms that break down organic material: those that need oxygen and those that don't. The ones that need oxygen don't smell, which is why mixing up your compost pile every two months or so (which aerates it with oxygen) is so important. If you don't mix your pile and it's too wet and compacted, the organisms that don't need oxygen will take over and a different kind of decomposition, called fermentation, will occur, giving off a foul odor and releasing methane.

- Fermentation is also the kind of decomposition that occurs when you send your organic garbage to a tightly packed landfill with little oxygen. In fact, organic materials rotting without oxygen in landfills account for almost *40 percent* of man-made methane emissions in the United States. Composting in your backyard, where there's plenty of oxygen, however, produces no methane emissions.

- A compost pile keeps decomposing in cold weather, only at a slightly slower rate. That's because the microorganisms that break down your compost generate their own heat as they work, meaning they can survive even in cold temperatures.

- ***Climate Results:*** There are 2,313 landfills in the United States. If every American composted his or her organic garbage this year, we would reduce the national waste stream by the size of 460 landfills.

- ***Money Matters:*** During the height of the autumn leaf-raking season, leaves make up 75 percent of our nation's total waste stream and cost the United States nearly one billion dollars in tax dollars to dispose of them. In other words, America would have one billion more dollars to spend on better causes if we all composted our leaves this year.

Easy Ways You Can Help

- ***Start a compost pile!*** Once you get the hang of it, you'll be amazed at how easy composting is. Here's a simple checklist to help you get started:

  1. ***The Spot.*** Pick a location for your compost pile that's convenient (not too far from your kitchen). If it's too far away, you won't use it!
  2. ***The Pile (or Bin).*** Choose a partially shaded, level spot either to start your compost pile or to place your composting bin (available at hardware stores). A simple pile will compost just as well as a bin, although you may

want to put a small fence around your pile to keep the wind from blowing it. Check to see if your city requires the use of a bin.

3. ***The Stuff.*** There are two types of composting materials: greens and browns. The microorganisms that do the decomposing need green, nitrogen-rich materials in order to grow and brown, carbon-rich materials for energy. The key to a good compost pile is always to use *both* green and brown materials, ideally in a one-part-green to three-parts-brown ratio. If you use only brown materials, your pile will take a very long time to decompose. If you use only green materials, your pile will attract flies and animals and give off a foul odor.

Include	Don't Include*
Three Parts Brown: Leaves, bark, wood chips, chopped brush, shredded newspaper, nonrecyclable paper, sawdust, fireplace ash, vacuum cleaner lint, wool and cotton rags.	Oil, grease, fat, meat, fish, dairy products, bones, peanut butter, mayonnaise, cat or dog manure (because it can spread certain diseases), insect-ridden or sickly plants.
One Part Green: Grass, yard trimmings, fruits and vegetables (and their peels!), bread, eggshells, coffee grounds *and* filters, tea bags, freshly picked weeds, houseplants, manure (cow, horse, pig, chicken, or rabbit)	**As long as you don't include these things, your compost pile will not attract insects or animals or give off foul odors.*

This list is not all-inclusive. Adapted from U.S. Composting Council, "What Goes in Your Compost Pile."

4. ***The How.*** To begin a compost pile, simply pile up any materials listed in the *Include* column in the preceding table. Don't let the pile get bigger than three cubic feet (start a second pile if you have to). When adding kitchen scraps, dig a hole in the compost pile, drop

them in, and cover them up to keep pests away. Every
two months or so, go out and turn or mix your pile with
a shovel to give it some oxygen. Besides that, there's
nothing you have to do. Just let it sit there!

5. ***The Results.*** After six months you will start to find finished compost (or humus) at the bottom of the pile,
ready to use on your lawn, garden, and houseplants. Collect the humus from the bottom of the pile and work it
into your garden soil or use it as a top-dressing.

- ***Rake those leaves.*** When the leaves fall in autumn, you'll
find yourself stuck with more brown materials than you can
handle. The solution is to create an all-brown compost pile
next to your actual compost pile. Remember, it's okay to have
an all-brown pile (it will just take a long time to decompose),
but an all-green pile will attract pests.

- ***Leave it a-lawn.*** Think putting your grass clippings into a
compost pile is easy? Leaving them on your lawn is even easier, and can save you money! See Tip 32 for more info.

- ***Compost in your community.*** Creating a compost pile in
your backyard is just one way to compost. Many cities have
community compost centers where you can drop off your
food scraps and yard trimmings. In that case, you can freeze
your food scraps in a container until you have a chance to
drop them off. In addition, over 2,000 cities have adopted a
pay-as-you-throw garbage collection system, which gives residents a financial incentive to compost and recycle. Call your
local government to see if your community composts—if not,
ask them to start a program.

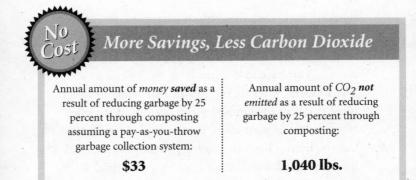

More Savings, Less Carbon Dioxide

No Cost

Annual amount of *money saved* as a result of reducing garbage by 25 percent through composting assuming a pay-as-you-throw garbage collection system:

$33

Annual amount of CO_2 *not emitted* as a result of reducing garbage by 25 percent through composting:

1,040 lbs.

Important Note: If you pay a flat rate for your garbage disposal but buy mulch or fertilizer for your lawn or garden, you can still save at least $25 a year by not having to buy as much mulch or fertilizer.

Assumptions: Based on a typical pay-as-you-throw garbage disposal system in which 32-gallon garbage bag costs $1.25; typical household puts out two bags per week; 1 pound of waste eliminated = 1 pound of carbon dioxide not emitted; a full 32-gallon garbage bag weighs approximately 40 pounds.

Search For More Info

- http://compostingcouncil.org/pdf/home_composting_faq. pdf—Here are easy-to-understand answers to all your questions about how to start a home compost pile.

- www.mastercomposter.com/pile/advbuild.html—Check out this step-by-step guide to building a compost pile.

- www.epa.gov/epaoswer/non-hw/muncpl/factbook/Internet/ redf/reduce2.htm—Go here to learn more about the science behind composting.

Recycling Is Here to Stay

In the time it takes you to read this sentence, Americans will buy, use, and throw away 1,245 tons of raw materials.

Overview. When we toss a soda can into a recycling bin, there's always that lingering doubt—Am I really making a difference? The answer is a resounding *yes*! It's true that hauling recyclables to a plant, cleaning them, melting them down, and remolding them into new products takes energy. But recycling takes far *less* energy than hauling all that garbage to faraway landfills and then using new raw materials from the Earth to manufacture brand-new products. Besides conserving energy and natural resources, recycling reduces the amount of landfill space needed—which means more room for parks and forests. It also reduces water pollution by stopping the chemicals in our garbage from seeping into the underground fresh-water supply. And perhaps the best thing about recycling is that it creates jobs—five times as many jobs as landfilling does. So, first reduce, then reuse, and then recycle, recycle, recycle!

What You Should Know

- Americans throw away 40,000 plastic bottles and 50,000 plastic bags every minute.

- Contrary to popular belief, recycling does *not* use more energy than manufacturing a brand-new product. Making recycled paper uses 55 to 75 percent less energy than making virgin paper, making things out of recycled plastic uses 60 to 75 percent less energy, and making products out of recycled aluminum uses 95 percent less energy!

- For every soda can you recycle, you'll prevent 1 pound of carbon dioxide from being emitted. If you recycle your newspaper every day, you'll prevent 100 pounds of carbon dioxide from being emitted each year.

- In the last 30 years the U.S. recycling rate has quadrupled, which would make you think that the amount of trash we now send to landfills has dropped by the same amount. Wrong. The amount of trash we send to landfills has actually gone *up* over the last 30 years—by 33 percent! In other words, even though we're recycling more than ever, we're also *throwing away* more than ever.

- Americans throw out 270 million automobile tires a year.

- Good state: Seattle, Washington, has set a goal to increase their recycling rate to 60 percent by 2008 (they are currently at 33 percent, while the national average is 28 percent). Bad state: Montana currently has the worst recycling rate in the country, recycling only 5 percent of its garbage.

- Americans use 187,000 tons of paper every day. For every ton of office paper we recycle, we prevent 17 trees from being cut down—and don't forget that by not being cut down, those 17 trees continue to soak up carbon dioxide (see Tip 31).

- By throwing away just two aluminum cans instead of recycling them, you waste more energy than a person living in a third world country uses in an entire day.

- Glass does not lose any of its durability or quality when it's recycled—that's why the U.S. government now requires all newly manufactured glass containers to contain at least 35 percent recycled glass.

- Each year 200 million gallons of used motor oil are dumped on the ground, sent to landfills (after being thrown in the garbage), or poured down the drain by Americans—that's enough oil to fill up 120 *Exxon Valdezes*! Furthermore, just 1 gallon of used motor oil is enough to pollute 1 *million* gallons of drinking water. By contrast, if we recycled all that used oil, we would be able to import 1.3 million *fewer* barrels of oil a day.

- With all the steel and tin cans Americans use in *one day*, you could build a steel pipe from Los Angeles to New York and back.

- ***Climate Results:*** If 100,000 people who didn't recycle *started* recycling, they would collectively prevent 42,000 *tons* of carbon dioxide from being emitted each year.

- ***Money Matters:*** If California meets its goal of a 50 percent recycling rate by 2008, they will have infused two billion dollars into the state economy and created more than 45,000 jobs.

Easy Ways You Can Help

- *Know your community.* Today nearly every community in the United States either provides a weekly curbside collection of recyclables or has designated drop-off or buyback centers. If you're unsure whether your community recycles, call your local government office today and find out. Ask what can be recycled, how, when, and if there are any rules (i.e., whether you can mix glass and plastic). You can also enter your zip code into this wonderful web site, www.earth911.org/master.asp?s=ls&a=Recycle&cat=1, or call the national recycling hotline at 1-800-CLEANUP (1-800-253-2687) to find out about your area's recycling options. If your community doesn't have a recycling program, ask that they start one!

- *Recycle the usual.* That means all of your *paper* (newspapers, cardboard, magazines, juice boxes, milk cartons, grocery bags, et cetera), *glass* (bottles and jars of *any* color glass), *steel and aluminum* (soda, beer, and soup cans and their caps or lids), and *plastic* (jugs and bottles, some bags). Get multiple smaller garbage cans to make separating your recyclables easier.

- *Recycle the not-so-usual.* Put your used *motor oil* into a clean milk jug, label it, and take it to your nearest oil change or auto parts store to be recycled. Give your old car *tires* to the store where you buy your new ones, and make sure they recycle them. You may have to pay a small fee, but at least you'll know they'll be used to make road asphalt or new tires instead of clogging up our landfills. Try to donate any *wood* you have or look up a local tree removal service and ask if they'll take it.

Batteries contain many toxic chemicals, such as mercury and cadmium, which are released in landfills when the batteries get crushed. Do *not* simply throw your batteries away. Most watch stores will take back your used button batteries for free and recycle them. Most service stations now are required to accept and recycle your used car batteries. To find the nearest drop-off location for depleted *rechargeable* batteries, call 1-800-8BATTERY (1-800-8-228-8379) or check this web site: www.rbrc.org/consumer/uslocate.html.

Take all your disposable batteries, oil-based *paint*, and any burned-out *lightbulbs* (which contain trace amounts of mercury) to your local household hazardous waste collection center. You can store these things in your closet or basement and simply make one trip every six months. Call your Department of Public Works to find the household hazardous waste center nearest you.

- **Recycle your coolant.** You probably know that chlorofluorocarbons (CFC's) eat away at the ozone layer, but did you know they're also potent greenhouse gases that cause global warming? Leaky home and car air conditioners (which contain CFC's in their coolant fluid) make up the largest single source of chlorofluorocarbons in the United States. That's because a typical car air conditioner leaks the equivalent of 200 pounds of carbon dioxide *every year*! Ask your serviceperson to check for leaks and to capture *and recycle* the coolant when you take your car in for its biannual tune-up or whenever you have your home air conditioner serviced.

- **Recycle the big stuff.** Steel can be remelted an infinite number of times without degrading its quality, which is why the automobile recycling industry is doing so well ($8.2 billion in annual sales and producing enough recycled steel to manufacture 95 percent of all new cars!). When you're ready to junk your *old car*, either donate it to a charity such as the National Kidney Foundation (1-800-488-CARS) and enjoy the huge tax write-off or sell it to a scrapyard—look in the yellow pages under "Auto Salvage," "Automobile Wrecking," "Salvage," or "Scrap Metals."

  Any *appliance* that contains chlorofluorocarbons or coolants (refrigerators, air conditioners, et cetera) needs to be drained by a professional before you can legally throw it away. That way the toxic coolant can be captured and recycled instead of leaking into our landfills (and then into our groundwater). If the appliance is still in working condition, donate it. If not, call your Sanitation Department to find out how to dispose of the appliance. Usually someone will come to your house (for free), drain the coolant, and then tag the appliance so you can put it on the curb for recycling.

  Nine million tons of asphalt *roofing shingles* are dumped into our nation's landfills each year, costing us $400 million in disposal costs. That's probably because most of us don't realize that these shingles can be recycled into road asphalt or new roofing shingles. When you're getting a new roof, call your Sanitation Department and ask them how and where you can drop off your old shingles for recycling. Or try searching this Web site, www.thebluebook.com, by typing in "recycling center construction debris" for your area.

No Cost

More Savings, Less Carbon Dioxide

Annual amount of *money **saved*** for *your community* in disposal costs as a result of recycling two more pounds of materials a week:

$10

Annual amount of CO_2 ***not** emitted* as a result of recycling two or more pounds of materials a week:

130 lbs.

*Note: If you don't recycle at all and start to (thereby recycling 13 pounds per week), you'll save your community $65 a year and personally prevent 850 pounds of carbon dioxide from being emitted each year. Also, if you live in California, Connecticut, Delaware, Iowa, Massachusetts, Maine, Michigan, New York, Oregon, or Vermont, you can actually earn money by collecting a bottle refund when you recycle glass or plastic bottles.

Assumptions: One ton recycled material = 1.24 tons of carbon dioxide prevented from being emitted and $187 saved for the community.

Search For More Info

- www.nyc.gov/html/dos/html/dispose/dispose_a.html—Want to know how to recycle random things? Check out this alphabetical list. (Even though it's specific to New York, you'll at least know how one *should* recycle these things.)

- www.epa.gov/epaoswer/non-hw/muncpl/mswdata.htm#item4 —Go here to find out your state's recycling rate. Scroll up and down this page to see other surprising recycling statistics.

- www.epa.gov/epaoswer/non-hw/recycle/reuse.htm—Check out the Environmental Protection Agency's quarterly *Reusable News* for the latest news on recycling.

Expand Your Impact

Buy Green Electricity

Electricity generation is responsible for more than one-third of the United States' total carbon dioxide emissions.

Overview. We're almost there. If our goal was to deplete the world's store of fossil fuels, which took 200 million years to form, in just 200 years, we've just about done it. Scientists estimate that anywhere from 50 to 150 years from now, solar panels won't be an option, they'll be a necessity—because that's when we're going to run out of fossil fuels. The transition to renewable power sources that have no global warming or air pollution emissions, such as wind, solar, or geothermal power, *will* happen—when there aren't any fossil fuels left. If we wait until then, however, the damage to our climate may have already been done.

What You Should Know

- Electricity that is made from renewable sources, such as sunlight or wind, is called green power. More and more utilities are giving their customers the option to buy green power instead of coal-powered electricity, especially in states that have deregulated their electricity markets.

- The World Bank estimates that total global electricity generation will increase from 3.2 million megawatts in 2000 to 5.0 million megawatts by 2020. At this rate, many experts predict fossil fuels will run out by 2060. The solution? Switch to renewable sources of electricity.

- In 1970 the United States imported 34 percent of its oil. In 2000 we had to import 54 percent. The government predicts that we will have to import up to 75 percent of our oil by the year 2010—unless we transition to renewable power, that is.

- *Climate Results:* If 100,000 households bought renewable electricity instead of fossil-fuel-powered electricity for just one month out of the year, they would prevent 70,000 *tons* of carbon dioxide from being emitted.

- *Money Matters:* The renewable power industry could infuse a significant amount of money into the United States economy, in contrast to the amount we currently spend importing foreign oil each year. For example, the United States is in a position to corner the soon-to-explode market of photovoltaic cells, which convert sunlight directly into electricity (see Tip 17). We currently make two-thirds of the world's photovoltaic cells, and they already bring in $300 million a year.

Easy Ways You Can Help

- *Know what's up.* We burn three fossil fuels—coal, natural gas, and oil (a.k.a. petroleum)—to provide us with electricity, transportation fuel, and heat. The burning of fossil fuels represents 98 percent of the United States' total carbon dioxide emissions, and the United States currently generates more than 90 percent of its electricity from fossil fuels. The pollution that results from the burning of fossil fuels has been proven to cause lung and heart disease, liver and central nervous system damage, birth defects, and cancer. It also causes acid rain, smog, and water contamination. In fact, 64,000 Americans die prematurely each year from heart and lung diseases caused by particulate air pollution.

- ***Know the truth about nuclear power.*** You may have heard that nuclear-powered electricity is "clean." Well, it *is* clean in the sense that nuclear power gives off no air pollution or carbon dioxide emissions. However, nuclear power plants create vast amounts of radioactive wastes, which must be stored for over *200,000 years* before they lose their radioactive properties. When accidents occur at nuclear power plants, like the one at the Chernobyl plant in Ukraine, millions of people get exposed to radiation, which can cause burns, chromosome mutation, and cancer. Nuclear power plants are also expensive to build and operate.

- ***Know the truth about natural gas.*** You may also have heard that natural gas is a "cleaner" fossil fuel. Although it does have much less air pollution emission than coal or oil, natural gas emits just as much carbon dioxide. And like coal and oil, it's a fossil fuel that's quickly running out. Natural gas is a good *transition* fuel as we move toward cleaner electricity, but renewable power needs to be our ultimate goal.

- ***Know your renewables.*** Here are the six major sources of renewable energy that you should know about, in our order of preference:

Energy Type	How It Works	Environmental Impact
Solar	Photovoltaic cells convert sunlight into electricity	Absolutely no emissions
Wind	Windmills are used to create electricity	Absolutely no emissions
Biomass	Plants are burned to create electricity	If done on-site and carefully, no emissions
Ocean	Waves and tides are used to create electricity	No emissions but expensive to build
Geothermal	Water from underground reservoirs is used to create electricity	Very low carbon dioxide emissions
Hydroelectric	Falling water from dams or waterfalls is used to create electricity	No emissions but damaging to fish and wildlife

Unlike the fossil fuels that are currently used to create electricity, renewable sources produce few to *no* global warming emissions, air and water pollution, or waste. More important, these sources will never run out. They can also be produced domestically, which means we wouldn't have to depend on Middle Eastern oil supplies, and we could create more jobs for our citizens (to run the renewable power plants).

- **Buy green power.** Your ability to buy green power will depend on what is available in your area. If you live in one of the states whose electricity market has been deregulated and is now competitive (as of press time: California, Connecticut, Illinois, Maine, Massachusetts, New Jersey, New York, Pennsylvania, Rhode Island, and Texas), you have the choice to purchase electricity from renewable sources instead of coal-powered companies (for more info, click on your state at: www.greene.org/your_e_ choices/pyp.html). Because the renewable industry is just getting started, you have to pay a small fee (premium) for using renewable power. The national average premium is three cents per kilowatt-hour. Soon though, the market will become more competitive and this fee will disappear. Go here, www.eia.doe.gov/cneaf/electricity/chg_str/regmap.html, to see if or when your state plans to deregulate its energy market.

If you don't live in a state that has deregulated its electricity market, you might live in one of 29 states where some of the regulated utilities have *voluntarily* chosen to offer renewable power to their customers. This is called green pricing. Again, because the renewable industry is just taking off, the cost of this power will be slightly more than that of your traditional electricity. To see if your utility offers a green pricing program, go here and scroll down to find your state: www.eren.doe.gov/greenpower/summary.shtml.

Last, if you live in a state that has not made any plans to deregulate *and* whose utilities have not voluntarily offered green power (such as Louisiana, North Carolina, and Vermont), there's still hope! First, pester your utility and ask them to offer green pricing voluntarily. Second, buy green energy certificates, which allow you to support the green power industry even if your state doesn't. Go here for more info: www.green-e.org/your_e_choices/trcs.html.

- *Are we crazy?* We know we're asking a lot. We're asking you to pay extra for a certain kind of electricity simply because it's better for the environment. And you're thinking, Will that really make a difference? *Yes.* If the coal power companies lose just one customer to a renewable power company, they'll notice. The only way the renewable power industry will ever overcome the monopoly of the fossil fuel industry is if we support it now, when it's trying to get off its feet.

More Savings, Less Carbon Dioxide

Annual amount of *money* **spent** as a result of spending one extra dollar a month on renewable electricity (through a green pricing program):

$12

Annual amount of CO_2 *not emitted* as a result of spending one extra dollar a month on renewable electricity (through a green pricing program):

650 lbs.

Assumptions: One kilowatt-hour of electricity produced by fossil fuels is responsible for 1.64 pounds of carbon dioxide emissions. Based on 3 cents per kilowatt-hour national average cost of green pricing premium.

Search For More Info

- www.epa.gov/greenpower/whatis/whatis.htm—Here is a concise introduction to green power written by the Environmental Protection Agency.

- www.powerscorecard.org/technologies.cfm—Here are some great explanations of the pros and cons of all the fossil fuel and renewable power sources. A must-read!

- www.eren.doe.gov/greenpower/home.shtml—This Department of Energy web site is *the* place to go for information on green power, pricing, and green energy certificates. If you live in a state that has a deregulated energy market, go here, www.greenmountain.com, to check out the nations' largest retailer of green power.

Invest in Green Stocks

The World Bank predicts that the solar electricity market will be worth four trillion dollars by 2030.

Overview. If you're investing money in the stock market anyway, why not invest it in a socially responsible manner? In other words, why invest in oil-drilling or carbon-dioxide-spewing factories when you could make just as much, if not more, money *and* help prevent global warming by investing in a wind power company? Not tempted yet? How about this—investing in photovoltaics or fuel cells now is like investing in plastics was during the 1970s. Simply put, the renewable energy industry is about to take off. These technologies are being refined and tested in laboratories as we speak, and they will be hitting the market *soon*. Fossil fuels will inevitably run out, which is why renewable energy *will be* the power of the future. By investing in it early on, you'll be helping the environment and your stock portfolio.

What You Should Know

- Many people are surprised to find out that socially responsible investment funds (SRI's) perform as well as, if not better than unscreened funds. According to Morningstar, an unbiased investment research firm, socially responsible funds saw the same double-digit returns that regular mutual funds experienced during the late 1990s and did no worse than their competitors when the market turned down in 2000. In addition, two months after the attacks on September 11, Morningstar reported that 40 percent of the SRI funds they track were outperforming 75 percent of their categories.

- Socially responsible investment funds can compete with non-screened funds because companies that care about the environment are often more energy-efficient than their competitors, giving them lower operating costs. This efficiency allows them to charge the same prices for their products or services while earning more profits.

- The World Energy Council predicts that renewable energy sources, not including hydroelectric power, will account for as much as 32 percent of the global supply of energy in 2050, up from 12 percent in 1990.

- Not only is wind power the fastest growing renewable energy source today, but it's also the cheapest. According to the World Energy Council, current wind energy prices are either equal to or less than those of electricity that's being produced by fossil fuels.

- *Money* magazine reports that widespread production of fuel cell passenger vehicles could occur within as little as 10 years (see Tip 39 for more on fuel cells).

- The Royal Dutch/Shell Group predicts that renewable energy sources will supply half the global energy demand by 2050.

- Approximately 80 million Americans, or 40 percent of our adult population, invest in the stock market each year. If each of these Americans switched just $50 of their current holdings to a share in a renewable energy company, they would infuse $4 billion into the renewable energy industry.

- *Money Matters:* Clean Edge, Inc., the leading market research firm for renewable energy, predicts that the clean vehicle market (meaning cars that use alternative fuels, hybrid-electric drives, or hydrogen fuel cells) will grow from $2 billion in 2000 to $48 billion by 2010.

Easy Ways You Can Help

- *Do your homework.* Socially responsible investment funds invest only in companies that are socially and/or environmentally responsible. By investing your money in these funds, you can be sure that you won't be supporting oil drillers, rainforest cutters, or tobacco sellers. There are currently more than 200 SRI's in the United States, and they collectively control more than two trillion dollars. Here's how to get started: First, to find out how green your current holdings are, go here: www.calvertgroup.com/sri_kwyo.asp. Second, research different SRI funds at the Social Investment Forum to decide in which one or more you want to invest. Start with their basic introduction to socially responsible investing at www.socialinvest.org/Areas/SRIGuide. Then, go here, www.socialinvest.org/areas/sriguide/mfpc.cfm, to see how over 100 of the biggest SRI funds have performed in the last month. Last, read their excellent answers to your most frequently asked questions about socially responsible investing here: www.socialinvest.org/Areas/News. Social Funds (www.socialfunds.com), is another great site to research and compare SRI's.

Once you've narrowed it down to one or two SRI's, you can get a free, unbiased ranking of them at www.morningstar.com. Last, be sure to take into account the funds' management fees. Socially responsible investment funds tend to have higher management fees than non-screened funds because they need to research both the performance *and* the social responsibility of each company. Socially responsible investment *index* funds tend to have lower fees than actively managed SRI *mutual* funds.

- **Invest in green funds.** Here are some of our top recommendations for SRI funds:

  Domini (www.domini.com) is one of the oldest and most established SRI funds in the United States, and they have recently made environmental criteria one of their top priorities. You can invest in their socially responsible index fund, bond fund, or money market account. Read more about some of the environmentally sound companies they invest in by going here: www.domini.com/Social-Screening/Environment/Company-Profiles/index.htm.

  The Calvert Group (www.calvertgroup.com) is an SRI fund that starts by determining the 1,000 largest companies in the United States. Of those companies, Calvert invests only in the ones that meet or exceed strict environmental and social standards. Go here to read more about their index fund: www.calvertgroup.com/funds_profile933.html.

We also highly recommend the Ariel (www.arielmutual funds. com), Citizens (www.efund.com), and Devcap (www.devcap.org) SRI funds.

- ***Invest in renewable energy companies.*** Because fossil fuels *will* eventually run out, the transition to renewable energy is inevitable. By investing in renewable energy companies now, not only are you helping to jump-start an industry but you're also getting in from the beginning—when the stocks are cheapest. Here are some of the biggest names in the various renewable energy sectors. Also, go here, www.cleanedge.com/ CEindex.php, to see today's 36 best-performing clean-technology companies.

Solar Power Companies

AstroPower, Del. www.astropower.com
BP Solar, Md. www.bpsolar.com
Kyocera, Japan
    www.kyocerasolar.com
Seimens, Germany
    www.siemens.com

Fuel Cell Companies

Avista, Wash. www.avistacorp.com
Ballard, Canada www.ballard.com
Millennium Cell, N.J.
    www.milleniumcell.com
Plug Power, N.Y.
    www.plugpower.com

Green Power Marketers (All U.S.)

Community Energy
    www.newwindenergy.com
EnergyGuide www.energyguide.com
Exelon Power Team www.pwrteam.com
Green Mountain
    www.greenmountain.com

Wind Power Companies

Colorado Wind, CO
    www.cogreenpower.org
Gamesa, Spain www.gamesa.es
Nordex, Germany
    www.nordex-online.com
Vestas, Denmark www.vestas.com

- ***Don't support the bad guys.*** Although oil and gasoline companies used to be the place to strike it rich, that's not necessarily the case anymore. Fossil fuels are fading fast. More important, it is the burning of these fossil fuels that is causing global warming. Even though some oil and gasoline companies have seen the light and begun to devote a lot of money and time into renewable energy research, we think it's greener of you to invest in the renewable energy companies just listed. Only when the oil and gasoline companies have *fully* switched to renewable energy do we recommend that you reinvest in them.

The same goes for automobile companies—we suggest you invest in fuel cell companies now, and reinvest in the automobile companies only when they've *completely* switched over to alternative fuels, hybrids, or fuel cells. That being said, we do wish to give high praise to Honda and Toyota for their major pushes in hybrid vehicles.

Investment
More Savings, Less Carbon Dioxide

Potential amount of *money **earned*** over 20 years as a result of investing $100 in a renewable energy company:	Potential amount of CO_2 ***not** emitted* over 20 years as a result of investing $100 in a renewable energy company:
$573	**13,455 lbs.**

Assumptions: Based on Morningstar reports showing that SRI's have done just as well as nonscreened funds in the long run; assume 10 percent per year earnings in either a nonscreened or an SRI fund; assume 20 pounds carbon dioxide emissions are prevented for every dollar invested in renewable energy instead of fossil fuels.

Search For More Info

- Don't know a thing about stocks? Go to the Motley Fool and click on "Fools School" to learn the basics of the stock market and investing: www.fool.com.

- Can I really do well by doing good? Scroll down to the fifth question at this site to find out why investing responsibly does not mean having to sacrifice profits: www.socialinvest.org/Areas/News. Also, explore Morningstar (www.morningstar.com) to find their most recent and objective articles on socially responsible investment funds. Simply scroll to the bottom of the home page and type "socially responsible" in the "search" box.

- Where can I learn more about SRI's? In addition to the many Internet sites listed in this chapter, The *Green Money Journal* (www.greenmoneyjournal.com) is an on-line magazine dedicated to SRI's.

Stay Current!

Knowledge is power.
— Francis Bacon (1561–1626)

Overview. We've been talking a lot about energy—how to conserve energy this way, how to reduce carbon dioxide emissions that way . . . in fact, *all* we've been talking about is energy! But now it's time to remember *why* we've been talking about it—because of that not so little thing called global warming. While it's important for you to replace your incandescent lightbulbs with compact fluorescent ones, it's just as important for you to understand *why* you're replacing them. That's why we've scoured the Internet to find the best web sites for you to stay up to date on climate change news. And even though this tip won't save you money or reduce your carbon dioxide emissions, we hope it will give you an incentive to keep vacuuming your refrigerator coils year after year.

Easy Ways You Can Help

- ***Master the basics.*** Don't know *exactly* what global warming is? Don't worry—you're not alone. If you're like most Americans, you have only a vague idea of how global warming happens. If you fit that description, flip to Appendix A, where we start from the beginning and explain everything having to do with global warming.

- **Know the news.** To see today's top climate change stories, as well as the top stories from the past two months, go here: http://dailynews.yahoo.com/fc/World/Global_Warming. Scroll down the left-hand side of the page and you'll see links to recent magazine and opinion articles about climate change. To find every article ever written about climate change, search these wonderful news archives: www.climateark.org/articles/2002/recent. For a sassy, easy-to-read overview of climate change news in recent months, check out the "This Just In" link on *Grist* magazine's "Heat Beat": www.gristmagazine.com/grist/heatbeat. Last, to get climate change headlines e-mailed to you every day for free, sign up here: http://envi-ronet.policy.net/warming/newsroom/index.vtml#2.

- **Become Kyoto savvy.** If you don't know what Kyoto means (that's okay!), or just want a basic overview and history of the policy issues regarding global warming, flip to Appendix B. After you've read that, stay up to date on global warming policy by going to the Weathervane by Resources for the Future: www.weathervane.rff.org. (Click on "Archives" first. Then try "News and Notes" for the most recent policy stories and "At the Negotiating Table" for an easy-to-read overview of the latest Kyoto-related negotiations.) Go here to sign up for their free, monthly e-mail announcement of new features and articles: www.rff.org/easy_instant_free.htm. Last, check out the Carbon Trader's headlines at www.thecarbontrader.com. Right now, most of their headlines deal with the Kyoto Protocol negotiations, but as the protocol begins to be implemented, these headlines will focus more on carbon emissions trading.

- ***Stay on top of studies.*** For a basic introduction to the science behind climate change, read Appendix A. Then, keep up with the latest scientific discoveries by checking in with the Union of Concerned Scientists: www.ucsusa.org (click on "Global Warming" under their list of "Issues," then click on any of the recent additions along the right-hand side of the page, especially those under the headings "Special Features" and "Climate Impacts"). Second, type in "climate change" on the search function of this NASA Internet site to pull up a list of the most recent global warming studies: http://pao.gsfc.nasa.gov. Last, check out this other NASA page, www.giss.nasa.gov/research/intro, to see a list of recent studies that have been "dumbed down" so that we normal people (nonscientists) can understand them.

- ***Read a free newsletter.*** There are many on-line newsletters out there that you can have e-mailed or sometimes even mailed to you for free that will give you a more in-depth look into climate change issues. Here is a great quarterly newsletter about climate change in general: www.esig.ucar.edu/newshp/index.html (click on "View the Newsletter in Portable Document Format [PDF]" to view the most current issue; click on the word "subscribe" to get it e-mailed or mailed to you for free; click on "Past Editions" to read previous issues).

  Second, here is a wonderful newsletter about how climate change is affecting our weather: www.ametsoc.org/AMS/

newsltr/index.html (check out the "Weather and Climate Briefs" section of each newsletter, and click on "Subscribe to Newsletter Listserve" to get a monthly e-mail reminder when a new issue goes on-line). Also, check out the Harvard Medical School quarterly review of three to four recent scientific discoveries relating to climate change: www.med.harvard.edu/chge/the-review.html (click on "Global Climate Change" to read the reviews; click on "Order an Online Subscription" to get the entire newsletter e-mailed to you).

Last, check out this easy-to-read quarterly newsletter recapping climate change policy in the United States and abroad: www.ccap.org (click on "Publications" to view the newsletters—unfortunately, they don't yet e-mail them to subscribers—you have to go to their web site to read them). *Eco* is a daily newsletter that's published only during the major climate change policy conferences, giving you (and the attending diplomats!) a day-by-day report on what's happening: www.climatenetwork.org/eco.

- *Join an on-line discussion group.* Go here, http://pacinst.org/cc_10.html, to read about the 15 or so top e-mail discussion groups on climate change. After reading the brief descriptions, pick the one that most interests you—then simply e-mail the list owner and join for free. Once you join, you'll start receiving all the e-mails the various members of the group send out, and you'll also be able to send e-mails to the whole group. This is a great way to discuss global warming and to hear what other people like yourself have to say about the issue.

- **Know where to go for help.** If you're trying to find out something specific about global warming and didn't find the answer in Appendix A or B, try searching for it at the Climate Ark, a wonderful search engine dedicated to climate change: www.climateark.org. If you still can't find the answer you're looking for, ask Dr. Global Change: http://gcrio.custhelp.com. If he can't help you, this page lists other places on the web where you can ask specific questions: www.gcrio.org/ask-doctor-links.shtml#I.

- **Hate to read things on a computer?** That's fine. Reading the daily newspaper or watching the evening news works just as well—whatever it takes to keep yourself updated on climate change. Also, we recommend checking out the following print sources on climate change:

  - *The Heat is On*, by Ross Gelbspan—Check out this thorough introduction to the science behind climate change, as well as the fossil fuel industry's attempts to downplay it.
  - *Stormy Weather* by Guy Dauncey—This book explains how communities and nations can reduce greenhouse gases.
  - *The Coming Storm*, by Mark Maslin. Check out this great book about the kinds of things global warming will do to our climate.
  - *The Change in the Weather*, by William K. Stevens. Read this fascinating introduction to global warming and how it is affecting our climate.
  - *State of the World 2002*, by the Worldwatch Institute. This book gives a detailed look at our world's most pressing problems, the main one being global warming.

- *The Consumer's Guide to Effective Environmental Choices*, by the Union of Concerned Scientists. This book describes some of the best ways you can fight for the environment.
- *The Official Earth Day Guide to Planet Repair*, by Denis Hayes. This book gives you even more ways you can be an environmentalist.
- *Making a Difference*, by Amy Irvine. This is an inspiring collection of true stories about concerned citizens in the United States who have made saving the environment their cause.
- *No-Regrets Remodeling: Creating a Comfortable, Healthy Home That Saves Energy*, a book by the producers of *Home Energy* magazine, gives in-depth information and instructions on how to conserve energy in your home.
- *Home Energy* magazine—This wonderful magazine offers in-depth, up-to-date info on how to conserve energy in your home. Subscribe at www.homeenergy.org.
- *Consumer Guide to Home Energy Savings*, seventh edition, by Alex Wilson et al. This concise book gives even more ways you can conserve energy in your home.
- *Save Energy, Save Money* by Alvin Ubell. To find out still more ways you can save energy and money in your home, check out this comprehensive book.
- "Home Savers: Tips on Saving Energy and Money at Home." Call 1-800-GET-PINK (1-800-438-7465) and ask for this free booklet written by the Department of Energy.

Important Note: Although some of these Internet sites may not be the most professional looking, they are, in our opinion, the best in terms of content. Believe us—we researched over 300 sites in writing this tip so that we could recommend only the best!

Get Involved

If you tell five people about what you learned from this book, and the next day each of them tells five new people, and so on, it will take only 14 days for the whole world to know.

Overview. So, you've completely overhauled your house, car, and lifestyle, and they're as energy-efficient as they're ever going to be. Your work here is done, right? Wrong. There are many more simple things you can do—like telling your friends to read this book or sending an e-mail to your congressional representative—that have the capability of making a much bigger impact than following the tips in this book. There's only one catch: They require some social interaction. C'mon, don't be shy. It's time to come out of your shell and tell everyone about your new energy-efficient ways!

Easy Ways You Can Help

- **Spread the word!** The easiest and best thing you can do to make more of an impact is simply to spread the word about this book. We wrote it because we wanted you to know that you can help yourself and the environment at the same time.

Tell your family and friends about what you've learned or about something that surprised you; better yet, lend them your copy of this book!

- ***Support green businesses.*** It's sad but true: Your money usually has more of an impact than your vote. Support businesses that use recycled products or paper (dry cleaners, banks, et cetera); use low-flow showerheads, sinks, or toilets (health clubs or public rest rooms); use alternative fuel vehicles (rental cars or airport-to-hotel shuttle buses); use energy-conscious watering techniques (golf courses); or use renewable energy. Ask business owners if they use these things. If they do, make sure they know that you're supporting them for that reason and ask them to advertise their actions so others will know too. If they don't use these things, urge them to!

- ***Ask nongreen businesses to change.*** Ask your local coffee shop or gas station to hang up a sign that encourages customers to bring in their own cups or thermoses. Ask the chefs of your favorite restaurants to put vegetarian dishes on the menu and cook with locally grown, organic foods. Also, ask them to donate leftover food each night to a local homeless shelter and compost the rest. Ask all local businesses to install bike racks, and ask your electric utility to start offering green power.

    Ask your grocery store manager to sell five-dollar cloth bags at the checkout. With the store's logo on them, they will be a free source of advertising! Also, ask the manager to start stocking more bulk items and fewer items with excess packaging (give specific examples). On that note, if you have an idea for how a specific company could reduce a product's packaging, send a letter to their marketing department—large corporations won't change their ways unless they have proof that their customers want them to change!

- *Be demanding of your local government.* If you want any of the following things for your town or city—a curbside recycling and/or composting program; a maximum garbage limit per home; a stricter lawn-watering or tree-preserving ordinance; carpool lanes and a carpool program; better and safer sidewalks, crosswalks, and bike lanes; or safer and more reliable public transportation that uses alternative fuels— there's only one thing you need to do. Go to your next local government meeting and speak up! Call ahead to get your item on the agenda and then say what you want, ask what you can do, and ask your local officials what *they* can do.

These people have been elected *and are getting paid* to serve you, so make them earn their keep! If they don't do anything, tell your local newspaper about their lack of action. The following two organizations can educate your community leaders about global warming and suggest actions that will help prevent global warming on a community scale without hurting the local economy:

- Climate Change Learning and Information Center: www.cclic.com
- International Council for Local Environmental Initiatives: www.iclei.org

- **Be demanding of your state and national government!**
  Now that you know how to keep yourself informed about
  what's going on with global warming policy (see Tip 50), you
  need to make sure your congressional representative does
  too. E-mail, call, or write to ask about his or her stance on
  global warming, what he or she has personally done about it,
  and what's being done about it in the U.S. government. Go
  to www.congress.org and type in your zip code to find all the
  contact information for your representative, senators, and the
  president. If you have a suggestion, tell them about it and ask
  them to get back to you on its progress. Again, if they ignore
  you, tell your local or state newspaper that they ignored you!

  In terms of global warming, some of the biggest national
  policy issues right now are signing on to the international
  treaty to reduce global warming emissions (called the Kyoto
  Protocol—See Appendix B), not drilling for oil in Alaska's
  Arctic National Wildlife Refuge, raising fuel-efficiency standards for airplanes and passenger vehicles (especially
  SUV's), and giving more funding to the renewable power,
  alternative fuel, and high-speed train industries.

- **Support an environmental group.** Maybe you don't have the
  time to go and march on Capitol Hill, or maybe you're just
  shy. Luckily, there are people who will do it for you! You can
  support them by supporting their organizations. A meager
  $10 contribution to Greenpeace each year goes a long way
  when 100,000 people make it. Take a few minutes to find an
  organization that fights for the things you care about most,

then give them what you can. Here are five of our favorites for the prevention of global warming:

- Natural Resources Defense Council: www.nrdc.org/glob alWarming/default.asp
- World Wildlife Fund: www.panda.org/climate
- Greenpeace: www.greenpeace.org
- Sierra Club: www.sierraclub.org/globalwarming
- Environmental Defense: www.environmentaldefense.org

- ***Write to us!*** We wrote this book because we thought you all should know that there are things you can do to prevent global warming and that you can take an active role in preserving the Earth for future generations while saving money. We gladly welcome any thoughts, ideas, or reactions that may have come to you from this book, and we will respond to and/or share them in the best way we can. You can e-mail us at jeff-and-kelly@preventglobalwarming.net. If applicable, we will put your ideas on our book's web site: www.prevent-globalwarming.net

The Final Tally

Whew! Congratulations, you made it through all 51 tips! Feeling overwhelmed? Are we asking you to do so much that you don't even know where to begin? We thought so. Even though each of our individual tips is easy, when taken as a whole they can seem *quite* intimidating. That's why we've written up this checklist to help you organize your thoughts.

If you follow *only the "No-Cost" and "Low-Cost"* tips in this book (the *single* tip listed under "More Savings, Less Carbon Dioxide" near the end of each chapter), not including the Tips 13, 46, 47, and 48, you will save approximately *$2,022 and 25,919 pounds of carbon dioxide every year!*

Remember that these numbers could be higher or lower depending on your specific home, lifestyle, or climate. We're confident that you can figure out a way to use an extra $2,000 a year, but you might be wondering what saving 25,919 pounds of carbon dioxide will really do for global warming. Well, let us try to put things into perspective:

- If 1 person follows the no- and low-cost tips in this book, he or she will save an amount of carbon dioxide equivalent to the amount that *518 trees soak up in a year.*

- If 10 people follow the no- and low-cost tips in this book, they will save an amount of carbon dioxide equivalent to burning 276 barrels of oil.

If 100 people follow the no- and low-cost tips in this book, they will save an amount of carbon dioxide equivalent to taking *216 cars off the road for an entire year*.

- If 100,000 people follow the no- and low-cost tips in this book, they will save an amount of carbon dioxide equivalent to *permanently retiring a 400-megawatt coal power plant.*

- If 95 million people (35 percent of the U.S. population) follow the no- and low-cost tips in this book, they will save an amount of carbon dioxide equivalent to *achieving the United States' original emissions reduction target under the Kyoto Protocol!*

That's right. If 35 percent of the U.S. population were to follow only the no- and low-cost tips in this book, they would reduce our national greenhouse gas emissions by the amount asked of the United States by the Kyoto Protocol (the international global warming treaty—see Appendix B). Furthermore, they would achieve this without harming the economy; in fact, the money saved by following these tips could actually *stimulate* the economy, since it would eventually be spent on other things.

Overall, the first thing you should do is focus on Tip 51: Get involved and spread the word! The more people who follow our tips, the bigger the difference will be for global warming. As for the rest of the tips, we've put together a suggested order for following the specific tip listed under "More Savings, Less Carbon Dioxide" near the end of each chapter. The basic concept is to follow the no-cost tips first, then use the savings to implement the low-cost tips, and finally, use those savings to follow the investment tips.

Suggested Order for Following the 51 Tips

(the single tip listed under "More Savings, Less Carbon Dioxide" near the end of each chapter)

Do every day (no cost, start right away)
9. Shake 'n' Bake—Efficiently
10. Now You're Cooking
34. Your Car: The Carbon-Dioxide-Spewing Monster
46. Compost Most Everything
47. Recycling Is Here to Stay
50. Stay Current
51. Get Involved

Do twice a week (no cost, start right away)
4. You're Such a Dish
11. Phantom Loads

Do once a week (no cost, start right away)
3. Don't Be a Wash-Out
35. Learn How to Drive Less
36. Let Someone Else Do the Driving
37. Use Person Power
41. Eat Your Broccoli
43. Less Is More

Do once a month (investment, start right away)
48. Buy Green Electricity

Do once a year (no-cost)
38. Trains Not Planes
42. If You Buy It, It Will Come

Do the first weekend (no cost)
2. Rein in the Fridge
7. Troubling Toilets
12. You've Got (E-)Mail

Do on a weekend one month later (low cost)
1. Light Up Your Life
5. Winch the Water Heater

Do on a weekend two months later (low cost)
8. Got Evian?

(Continued on next page.)

Do on a weekend at the beginning of summer (no or low cost)

20. Assess the A/C
22. Dress Up Your Windows
33. Get Smart About Sprinklers

Do once or twice a week, all summer long (no cost)

32. Green Plants, Less Water

Do once in the middle of summer (no or low cost)

40. Down Home Cookin'
45. Be Thrifty

Do the weekend after the yard sale (no or low cost)

6. Shower Simply
13. Things Not Everyone Has
44. Use It Again, Sam

Do in the beginning of winter (no or low cost)

18. Fun with Furnaces
19. Do the Right Temp
23. Plug Air Leaks
25. Insulate Your Home

Do in the middle of winter (investment)

14. Out with the Old, Save with the New
16. Upgrade Your Water Heater

Do in the beginning of second summer (low cost)

21. Alternative Ways to Cool
24. Air Out the Attic
30. Save with Shade
31. Tree: Nature's Air Conditioners

After that . . . make the investments

15. All I Want for Christmas Is a New Laptop
17. Switch to Solar
26. Leaky Ducts
27. Replace Your Windows
28. Invest in a New Furnace or Air Conditioner
29. The Advantages of an Audit
39. New Cars, New Fuels
49. Invest in Green Stocks

Appendixes

Appendix A: The Science Behind Global Warming

1. What Is Global Warming?

Greenhouse Effect = Good

The Earth's atmosphere is made up of many gases, including nitrogen, oxygen, carbon dioxide, and methane. Some of these gases, such as carbon dioxide and methane, have the special ability to trap the sun's heat inside the Earth's atmosphere. This is why they're called greenhouse gases. In a greenhouse, where you grow plants, the sun's heat first enters through the greenhouse windows, and then the glass of the windows traps most of the heat inside, making it unseasonably warm and therefore ideal for growing plants all year round.

The greenhouse effect occurs naturally in our atmosphere. It's *supposed* to happen. Without these greenhouse gases trapping 70 percent of the sun's heat, the Earth's average temperature would be around 0°F, which is not conducive to farming and ultimately inhabitable for humans. Instead, the greenhouse gases keep the Earth at its cozy, average temperature of 57°F.

Global Warming = Bad

Since the industrial revolution of the mid-1700s, humans have been burning increasingly larger amounts of coal, oil, gasoline, and other fossil fuels in order to heat our homes, produce electricity, and power our vehicles. Unfortunately, when any of these fuels are burned, they emit the greenhouse gas carbon dioxide, as well as many toxic air pollutants.

By burning so many fossil fuels, we have been spewing an inordinate amount of carbon dioxide and other greenhouse gases into the air. As a result, we currently have an overabundance of greenhouse gases in our atmosphere, which means they are trapping more of the sun's heat than normal. Consequently, the Earth's average temperature is slowly, but surely, rising. This is what we call global warming, or climate change.

2. What Evidence Proves That Global Warming Is Happening?

Ever since a Swedish scientist named Svante Arrhenius predicted in 1896 that rising levels of carbon dioxide would eventually impact our climate, scientists around the world have debated global warming. When global temperatures rose sharply in the 1970s, scientists from around the world decided to stop arguing and start working together. As a result, in 1988, the United Nation's Intergovernmental Panel on Climate Change (IPCC) was formed. Two thousand of the world's top scientists from over 30 nations were given the difficult task of collecting comprehensive, unbiased data on climate change so it could be used to formulate government policies. Here is a brief summary of what they have found since 1988.

Temperatures Are Rising

Since 1988, the IPCC has issued four reports on climate change. Their most recent report, issued in 2007, states that over the last century, the Earth's average temperature has risen by 1.3°F and is predicted to rise another 3.2° to 7.2°F by 2100 (the range reflects the IPCC's best and worst case scenarios). Some areas of the world have already warmed by 5°F. According to the IPCC's latest report, 11 of the last 12 years (1995–2006) are among the 12 warmest years ever recorded. The IPCC's figures and predictions are disconcerting considering the fact that, since the dawn of human civilization 11,000 years ago, the climate has remained at a steady average of 57°F. Before then, the Earth experienced many climate changes, from ice ages to swamp eras. However, for a reason unknown to our scientists, the Earth's climate suddenly stabilized 11,000 years ago at 57°F, and it is this temperature *stability* that has allowed the human race to flourish. So, the first reason to be concerned is simply the fact that the climate is changing *at all*.

The second reason to be worried is that the climate is predicted to change by *so much*. Now, at first glance, a 1.3°F rise in temperature, or a predicted 7.2°F rise over the next century, may not seem like much. However, when you consider that it only takes an 11°F decrease in average temperature to send the Earth into a 70,000-year-long ice age, 1.3°F suddenly seems much more significant. It is difficult to predict exactly what will happen when the average temperature *increases* by 11°F, but the last time that happened (over 50 million years ago), the ice caps completely melted, England was a tropical swamp, and there were alligators near the North Pole. Some scientists think we're actually headed for another *ice age*, since prior to each of the last four ice ages, the average temperature first rose 3.6° to 5.4°F and then plunged sharply. If this theory is correct, we could very well see the onset of another ice age by 2100. One thing our scientists know for certain is that drastic climate changes can occur within *decades*. In summary, it is almost certain that our children—if not we ourselves—will see dramatic and perhaps irreversible changes in our climate.

Carbon Dioxide Levels Are Rising

Along with this rise in global temperature, the IPCC reports a dramatic rise in greenhouse, especially carbon dioxide, gases over the past 250 years. The IPCC believes that this increase in greenhouse gases, with their ability to trap heat, is fueling global warming. This isn't the first time in the Earth's history that carbon dioxide levels have risen or fallen along with the temperature. In the last 400,000 years, there have been four times when the level of carbon dioxide in the atmosphere has fallen below 200 parts per million—and those were during the last four ice ages. Yet, never during those 400,000 years did the level of carbon dioxide rise above 298 parts per million. In the last 100 years, however, carbon dioxide levels have risen by 35 percent, from 280 parts per million to 379 parts per million! By examining ice cores, the IPCC has determined that today's carbon dioxide levels are the highest they've been in 650,000 years, and predicts that these levels will rise even more dramatically, to 540 to 970 parts per milloin, by the year 2100.

The levels of the other greenhouse gases (methane, nitrous oxides, chlorofluoroccarbons, etc.) have also increased since the beginning of the Industrial Revolution. However, there is currently more carbon dioxide in our atmosphere than all of the other greenhouse gases combined, which is why we chose to focus on carbon dioxide emissions in this book. Another troubling fact about carbon dioxide and the rest of the greenhouse gases is that they have extremely long lives. In other words, once carbon dioxide is emitted into the atmosphere (by the burning of fossil fuels, for instance), it continues to trap the sun's heat for over 100 years. That means that even if we completely stopped burning fossil fuels today, the greenhouse gases already in the atmosphere would keep contributing to global warming for another century.

Snow and Glaciers Are Melting

The IPCC reports that widespread decreases in glaciers and ice caps have contributed to the 6.6 inch rise in sea level over the past century. Also, snow cover in the Nothern Hemisphere decreased by an average of 5 percent *per year* during the late 1980s, and has decreased overall since 1966. They also report that, over the last century, the annual amount of time that lakes and rivers in the Northern Hemisphere have been covered by ice each winter has decreased by 12.3 days. According to the IPCC, there has also been a widespread shrinkage of mountain glaciers over the last century—at the current melting rate, all of the glaciers in Glacier National Park, Montana, will be gone by the year 2070. The IPCC also reports a 22 percent decrease in the amount of Arctic summer sea ice since 1978. Meanwhile, in the Southern Hemisphere, Antarctic ice shelves are melting and breaking off at an alarming rate. More than 5,400 square miles of ice—roughly the size of Connecticut and Rhode Island combined—have broken off since 1995.

As for the future, the IPCC predicts that we will see faster and earlier thaws each spring, which will disrupt agriculture and freshwater supplies. They expect Northern Hemisphere snow cover and sea ice to continue to decrease, and glaciers and polar ice caps to melt even further. It is the melting of glaciers and ice caps that will have the greatest impact on rising sea levels.

Humans Are the Cause

Perhaps the most important sentence in the IPCC's fourth report was: "Global atmospheric concentrations of carbon dioxide, methane, and nitrous oxide have increased markedly *as a result of human activities* since 1750." They go on to specify that the increase in carbon dioxide is "due primarily to fossil fuel use and land use change." Humans aren't the only ones who emit carbon dioxide—nature does too, when plants and animals die. However, these plants and animals absorb and store the same amount of carbon dioxide when they are alive, so that the *net* gain in carbon dioxide is zero. Nature works with the Earth in a balanced cycle; humans do not.

Among humans, the United States emits more greenhouse gases than any other country in the world—by a long shot. Even though we only make up less than 5 percent of the world's population, we emit 25 percent of the world's total greenhouse gas emissions. The next most culpable country, China, emits only 12 percent, half of what we do. Being the wealthiest country in the world does not give us the right to harm a planet that we share with 6 billion other people and that *all* of our children will inherit.

3. What Effects Will Global Warming Have Now and in the Future?

Warmer temperatures may sound appealing to you if you live in Minnesota, but unfortunately, a warmer winter is just *one* of the things global warming will bring (not to mention that warmer winters negatively affect agriculture crops, water supply, and wildlife). Because humans have never interfered with the climate as much as we have during the last 200 years, it is difficult for scientists to predict *exactly* what our actions will cause. However, they know that the following five trends are *already* occurring, and predict that they will only intensify as a result of global warming:

More Severe Weather

You might be asking yourself, What can a 1°F rise in average temperature *really* do? Well, for starters, it's already meant more severe and frequent heat waves, droughts, floods, cold spells, and storms. The unpredictable temperature swings of the past few years, the growing intensity of the El Niño effect, and the recent deadly heat waves are all *current* effects of global warming.

The years 2005 and 1998 are the two warmest on record. In 1998, Chicago saw its hottest day in history, Boston experienced record-breaking flash floods, South Dakota saw record snowfalls, and the Eastern United States had its driest growing season on record. Furthermore, the IPCC reports that, over the twentieth century, precipitation in the Northern Hemisphere increased by 5 to 10 percent, mostly as a result of heavier downpours. So, if you're wondering why the weather has been so crazy lately, the answer is global warming.

It's not going to stop there, however. Within the next 100 years, the IPCC predicts global warming will cause even deadlier flash floods, monsoons, droughts, heat waves, cold spells, and storms. These effects won't be universal, though. Some areas will experience flooding, while others will experience droughts; some will experience heat waves while others will experience unseasonable cold spells.

Human Health Impacts

The warmer global temperature will increase the rate of evaporation, which will decrease moisture levels in the soil and cause lake and river levels to drop, negatively affecting our agriculture industry and water supplies. The year 1999 had the driest growing season on record, causing agricultural emergencies to be declared in 15 states. In the spring of 2002, water restrictions were imposed on New York City for the first time in over a decade, as a result of a *winter* drought. In addition, deserts are expected to expand in area, encroaching on farmland. Warmer ocean temperatures have also already affected our fishing industry, such as when the population of Pacific salmon dropped sharply in 1998 as a result of overly warm water temperatures, causing fisheries to temporarily shut down.

We will see increased deaths due to heat waves, floods, and storms. In 2003, more than 35,000 people died in Europe from the deadliest heat wave on record. Air pollution will worsen as temperatures rise, increasing allergies, asthma attacks, smog, and acid rain for all regions. The drier soils will increase the chances of wildfires, which can threaten human homes and lives as well as natural habitats. In 2007, Southern California experienced its driest year on record, fueling wildfires that burned 500,000 acres of land and destroyed 1,500 homes.

One of the scariest prospects of global warming is the spread of tropical diseases. As the planet warms, the areas where disease-carrying mosquitoes can survive will expand. In Mexico, dengue fever has already spread to 5,600 feet above sea level—an elevation that used to be too cold for mosquitoes. In the last two decades, local malaria outbreaks have occurred in Michigan, New Jersey, New York, Texas, Florida, Georgia, and California. In 1999, the fatal

West Nile virus arrived for the first time in the United States and is now spreading rapidly. Other diseases that scientists expect will spread as a result of warmer temperatures are yellow fever, encephalitis, cholera, *E. coli*, Lyme disease, and salmonella. Scientists predict that, by the end of this century, we will be seeing 50 to 80 million additional cases of malaria per year compared to what we see now.

Flooding of Coastal Areas

According to the IPCC's fourth report, the global average sea level has already risen 6.6 inches over the last century as a result of glaciers melting into the ocean. That's three times faster than the historical rising rate. Florida farmland that is 1,000 feet inland is already being infiltrated by salty ocean water, which is poisoning the crops. Rising sea levels have already flooded low-lying areas such as Bangladesh, leaving 30 million people homeless, and have reduced marsh area in Maryland's Chesapeake Bay.

The IPCC predicts that sea levels will rise another 7 to 23 inches over the next century (best and worst case scenarios), which would flood island and coastal communities around the world, currently home to 50 to 70 percent of the world's population. Even a 17-inch rise in sea levels would erode 100 feet off U.S. beaches, flood 9,800 acres of Massachusetts coastal property, and 770 square miles of dry Florida land. Of particular vulnerability to flooding as a result of sea levels rising are the eastern and western coasts of the United States (especially Florida and the Mississippi Delta region), Western European coasts, and island groups such as Indonesia, Hawaii, and the Philippines. The Environmental Protection Agency reports that sea levels are rising faster along U.S. coasts than anywhere else in the world, and they predict that sea levels will rise one *foot* on the Eastern and Gulf coasts by as early as 2025. Major U.S. cities near sea level include New York City, Los Angeles, San Francisco, and Miami.

Economic Impacts

The real estate, construction, recreation, and tourism industries will all be negatively affected by these changes, but perhaps none will be as hard hit as the insurance industry. As a result of increased severe weather events, insurance rates will have to skyrocket in order for the insurance companies to stay in business. Damages from severe weather events in the 1990s, the warmest decade on record, cost insurance companies three times more than damages in the 1980s. Furthermore, poorer nations will be hardest hit by the effects of global warming, since they are more dependent on agriculture, have less stable water supplies, and lack the financial resources to prepare for and deal with the consequences of global warming.

Animals and Plants Will Become Extinct

Perhaps one of the saddest consequences of global warming will be the extinction of helpless plants and animals who cannot adapt quickly enough to the higher temperatures, increased precipitation, and drier soils caused by our burning of fossil fuels. The first to die off will most likely be the polar bears, who use the Arctic summer sea ice to hunt for food. As this sea ice breaks off earlier and earlier each year, the polar bears have less time to hunt for food. Scientists at the University of Washington predict that the Arctic summer sea ice will completely disappear by 2040. The loss of sea ice will also negatively alter the habitat of ring seals, beluga whales, penguins, and sea birds. In the fall of 2007, thousands of walruses were forced onto Alaska's beaches, since the Arctic sea ice that they usually inhabit had melted to a record low.

Caribou in Canada have experienced major die-offs as a result of heavier than normal snowfalls that cover up their food supply. The sea bird population in California has dramatically declined as a result of the warming of the California ocean currents, and certain West Coast butterflies are disappearing in the southern limits of their range. The extinction of more than 20 species of frogs in the Costa Rican Cloud Forest has been linked to warmer Pacific Ocean temperatures, and coral reefs are dying in the warmer than usual water. Plant and animal ranges will shift northward and upward as temperatures rise, as evidenced by the invasion of subalpine trees into rare, alpine meadow habitats. Eventually, plants and animals that live in the mountains will have nowhere else to go—except toward extinction. Similarly, as sea levels rise, plants that thrive in the tropics will be flooded out with salt water, such as the currently dying mangrove forests in Bermuda. Last, warmer temperatures will allow agricultural pests to thrive, just as they did in 2000 when warm, wet weather caused Sydney, Australia, to experience its worst outbreak of black locusts in 50 years.

4. What's the Deal with Fossil Fuels?

Hundreds of millions of years ago, long before the dinosaurs, the Earth was covered by swamps, bogs, and strange-looking plants and animals. When these living things died, their bodies decomposed. Over time, the remains of these plants and animals were covered by layer upon layer of mud, rock, and sand. After millions of years of decomposing, these remains eventually turned into today's fossil fuels: coal, natural gas, and oil (also known as petroleum or crude oil). Gasoline and diesel fuels are derived from oil. The kind of fuel that forms depends on the type of plant or animal that decomposed, the amount of time the remains were buried in the earth, and the temperature and pressure conditions that existed while they decomposed. Fossil fuels are nonrenewable in the sense that, once we deplete the current resources, they will not form again for hundreds of millions of years.

Fossil Fuels = Global Warming

Today, we burn fossil fuels to provide us with approximately 90 percent of our energy, which we use to heat our homes, drive our vehicles, and provide electricity (the other 10 percent comes from sources such as wood, charcoal, nuclear power, hydroelectricity, and renewables). The problem is, whenever we burn fossil fuels, the carbon that was buried deep in the Earth for hundreds of millions of years is released back into the air, where it mixes with oxygen to form carbon dioxide, a global warming gas. Global warming and all its consequences should be enough to persuade us to switch from fossil fuels to other sources of energy that don't harm the environment (such as solar or wind power). However, if you need a little more persuading, let us reiterate the fact that fossil fuels are *non*renewable . . .

Oil Is Running Out

At the world's current rate of consumption, we have enough coal to last us another 200 years (remember, though, that every time we burn coal or any other fossil fuel, we release carbon dioxide). The amount of oil that's left, however, is a completely different story. Oil currently provides 40 percent of our annual energy needs—that's more than any other fossil fuel (coal is next at 26 percent). To further complicate things, the global demand for oil is growing by the second. The consensus among scientists from around the world is that we will have depleted half of the world's total oil supply (in other words, global oil production will peak) sometime before 2018. They also agree that the half that we have just about depleted was relatively easy to find—the second half of the world's oil reserves will be much more difficult and expensive to extract.

It doesn't really matter when we *run out* of oil, though. The date that really matters is when we hit the *halfway* mark. The second we've depleted half of the Earth's oil (sometime before 2018), the world's demand for oil will outweigh the Earth's supply of oil, and prices will go through the roof as nations haggle for the last remaining barrels. Switching to coal is not an option, since we need *oil* to run our vehicles. Switching to natural gas is not an environmentally sound option (over a 20-year period, it is 9 percent worse than oil in terms of global warming), and besides, scientists predict that natural gas will reach its halfway mark within the next 15 to 30 years. If the thought of skyrocketing gasoline and natural gas prices isn't enough to convince you that we need to switch to renewable fuels, and fast, then let us give you one final reason . . .

Dependence on Foreign Oil

The United States currently imports over half of its oil, mostly from Middle Eastern countries. The Department of Energy estimates that we will import *60 percent* of our oil by the year 2030. Importing oil means not only giving taxpayers' money to other countries (about

$200,000 every minute!), but also having to do business with nations that don't always share the same ideas about human rights or environmental protection that we do. Last, since the events of September 11, 2001, national security has become one of our nation's top priorities. Flammable natural gas pipelines and nuclear power plants are considered prime targets, whereas solar panels and windmills carry no such risk (for more info on why nuclear power is not a good alternative to fossil fuels, see Tip 48).

What About Alaska?

Why not solve the problem of our dependence on foreign oil by drilling for oil in Alaska? you might ask. Well, there are many reasons. First of all, scientists from both sides of the debate predict that, even if we started drilling in the Alaska National Wildlife Refuge today, the oil would not reach the U.S. market for another 7 to 12 years. Furthermore, scientists predict that there are roughly 10.3 billion barrels of oil actually *recoverable* from the refuge over the next 50 years. Although that may sound like a lot, when you consider that the United States currently consumes 19.4 million barrels of oil *every day* (and that number is only expected to increase), the Alaska refuge will only provide us with a total of $1 1/2$ years' worth of oil over 50 years.

There are many other reasons why drilling for oil in Alaska is not a good idea. First of all, the refuge is home to over 130 species of birds, bears, rare oxen, caribou, polar bears, and other wildlife. Drilling for oil would significantly degrade their habitat. Furthermore, 95 percent of Alaska's Northern Slope is *already* available for oil drilling—in fact, it's been nearly depleted. Squeezing $1 1/2$ years' worth of oil out of the last 5 percent is not worth the destruction that it will cause to nature. A much smarter solution for our economy, national security, *and* the environment would be to invest in domestic alternative fuel technologies that don't contribute to global warming. For example, if all our cars ran on ethanol, which is made from corn, we could produce all our fuel from midwestern farm fields. If all our houses used energy from wind and solar power plants, we could forget about oil and get our electricity from the windy Plains States and the sun-soaked southern states. Solutions such as these are the only real, *long-term* solutions to reduce our dependence on foreign oil imports.

5. Other Factors to Consider

Aerosols: The Opposite of Carbon Dioxide?

Just as greenhouse gases have the ability to trap the sun's heat inside the Earth's atmosphere, some gases have the ability to reflect the sun's heat before it enters the Earth's atmosphere. These gases are called aerosols. Aerosols are microscopic particles or liquid droplets suspended in the air. They are caused by the burning of fossil fuels, biomass, and by the eruption of volcanoes. Since they block the sun's heat, aerosols have a cooling effect on the Earth, opposite to the global warming effect. However, this does not mean that they can balance out global warming. First of all, when fossil fuels are burned, much more carbon dioxide is released than aerosols, resulting in a net warming effect. Second, aerosols last in the atmosphere only for a few weeks, whereas one molecule of carbon dioxide will trap the sun's heat for more than a century. Last, and most important, aerosols such as soot and sulfur dioxide (SO_2) are the main causes of air pollution and acid rain. Some are even proven to cause cancer in humans. For our own health, we need to stop emitting aerosols just as desperately as we need to stop emitting greenhouse gases. (Note: The IPCC does take into account the cooling effect of aerosols in its calculations.)

Carbon Sinks . . .

You may have heard the term *carbon sink* in relation to global warming. Carbon sinks have to do with what we explained in Tip 31 about trees and carbon dioxide. To summarize, trees "breathe" in the opposite way that humans do. Humans breathe in oxygen and exhale carbon dioxide (called respiration). Plants, on the other hand, "breathe" in carbon dioxide from the air and release oxygen (called transpiration). They do this by first breaking down the carbon dioxide into carbon and oxygen, then storing the carbon inside their cells and releasing the unnecessary (for them) oxygen. When plants die and decompose, the carbon that had been accumulating inside their cells all their lives gets released back into the air, where it immediately bonds with oxygen to form carbon dioxide. Therefore, although plants take carbon dioxide out of the atmosphere and store it during their lifetime, they release it all back into the atmosphere when they die. This process of storing carbon is called carbon sequestration. Due to their ability to store or sequester large amounts of carbon, forests and other large areas of foliage have come to be known as carbon sinks.

. . . Are Only Temporary

When talking about possible solutions to global warming, it is essential that we understand that plants store carbon dioxide only *temporarily*. Many government leaders think that planting more trees will balance out our global warming emissions, because the trees will soak up all the carbon dioxide we emit. This is true—as long as we plant enough trees and as long as we make sure that, when those trees die, they are immediately replaced by new trees. However, there are many problems with this idea. First of all, the United States alone would have to cover 58 percent of its land with forest in order to absorb its current emissions from fossil fuels. With the world population currently around 6 billion and rising, there simply isn't enough room for all the forests we would need to soak up our emissions.

Second, the effects of global warming have already begun, and among them are increased forest fires and pest infestation. If wildfires were to consume a forest, all of the carbon that had been being stored in its trees would be instantly released back into the atmosphere, causing a sudden surge in carbon dioxide emissions. In addition, after the fire was over, we wouldn't be able to plant new trees for some time due to the charred ground. So much for our carbon sink.

The most important reason why carbon sinks are not a complete solution to global warming is that they are simply an excuse to keep burning fossil fuels. The only way to truly reduce greenhouse gas emissions is to *reduce the amount of fossil fuels we burn.* It's very simple! In order to make up for the damage that's already been done, we should plant more trees. In order to stop doing *more* damage, we need to dramatically reduce the amount of fossil fuels we burn in the first place.

6. So, What Should We Do?

The debate is over. Scientists conclusively agree that the planet is warming and that the burning of fossil fuels by humans is to blame.

The debate now is about what the exact effects of global warming will be and how we should go about trying to prevent them. There are many people, especially those who have a stake in the fossil fuel industry, who claim we should wait and see what happens, and if it's bad, only *then* should we do something about it. The problem with this kind of reasoning is, what if it's so bad we can't do anything to fix it? What if we're currently setting into motion an irreversible chain of major climactic changes that will threaten human existence, as many of the world's top scientists predict? In that case, some Noah's Ark type of preventative action sure seems like a good idea, doesn't it? And if the scientists turn out to be wrong, will we have really done any harm? *No.* As we've tried to point out in this book, preventing global warming can be *easy* while causing zero harm to the economy—in fact, it can even bolster it. With careful planning by our government, those who work in the fossil fuel industry can gradually be transitioned to the renewable energy industry. The key is to act sooner rather than later.

What You *Can Do*

Our task is simple: reduce greenhouse gas emissions. The IPCC's report clearly states that reductions in greenhouse gas emissions will be necessary to stabilize their warming effect on the Earth. *You* can help reduce greenhouse gas emissions by following the tips in this book. Remember, every time you use less electricity, gasoline, or natural gas, you're using less fossil fuels, which means fewer greenhouse gases are emitted. Our *government* can help reduce greenhouse gas emissions by signing on to an international treaty that promises to reduce our national greenhouse gas emissions (see Appendix B for more info).

7. Where Can I Go to Learn More?

To find other basic introductions to global warming, go to these web sites:

www.epa.gov/climatechange/index.html—This web site from the Environmental Protection Agency is loaded with easy-to-understand information on every aspect of global warming.

www.nrdc.org/globalWarming/f101.asp—Here are some thorough answers to basic questions about global warming.

www.climatecrisis.net—This is Al Gore's site about global warming. It's packed with excellent videos and graphs of the global warming crisis.

For more in-depth information about global warming, check out these books:

Global Warming: The Complete Briefing by John Houghton —An in-depth, comprehensive background on global warming and all its implications.

The Ice Chronicles by Paul Mayewski, Frank White, and Lynn Marqulis—The scientist who led the drilling of glaciers in Greenland explains how his findings relate to global warming.

Last, to stay on top of new scientific discoveries relating to global warming, see Tip 50.

Appendix B: What Governments Are Doing About Global Warming

The Framework Convention on Climate Change

When the Intergovernmental Panel on Climate Change (IPCC) came out with its first report in 1990, the findings were alarming enough that the world's leaders called for an international treaty to prevent global warming. In response, the United Nations presented a Framework Convention on Climate Change at the Earth Summit in Rio de Janeiro, Brazil, in 1992. The framework declared that the world's nations needed to band together to stabilize their greenhouse gas emissions to levels that would prevent dangerous, human-induced climate change. It further stated that they needed to stabilize these gases in a time frame that would not damage economic development, global ecosystems, or food production. The Framework Convention on Climate Change has been signed and ratified by over 180 countries, including the United States.

Conference of Parties

Since 1992 all the countries that ratified the framework—called "parties to the Convention"—have been meeting every one to two years to decide on the specifics of the framework. These meetings are called Conferences of the Parties (COP); the first conference was called COP1, the next COP2, and so on. After the first few COP's, the parties realized that if they truly wanted to reduce greenhouse gas emissions, they needed to add a legally binding agreement (called a protocol) to the framework, so that nations would be *required* to reduce their emissions or held accountable by law.

The Kyoto Protocol

The parties presented a first draft of this protocol at COP3, held in Kyoto, Japan, in 1997. For that reason, it was called the Kyoto Protocol. The Kyoto Protocol stated that 38 industrialized countries, including the United States, needed to reduce their greenhouse gas emissions to an average of 5.2 percent below their levels of 1990. The protocol further stated that these countries must meet their reduction targets by the year 2012. Specifically, it asked the United States to reduce its emissions to 7 percent below their 1990 emission levels.

United States' Withdrawal from Kyoto

One of the first things the Bush administration did after President Bush took office in 2001 was reject the Kyoto Protocol and withdraw the United States from all formal negotiations related to it. The administration's reasoning was that meeting targets would be too costly for the U.S. economy. Ironically, soon after this rejection, the parties finalized and approved an emissions trading system that would make meeting emission reduction targets much less costly. A year after rejecting the protocol, President Bush announced an alternative plan. His initiative called for steady *increases* in greenhouse gas emissions instead of reductions. A storm of criticism ensued, and the "Clear Skies Initiative" never took flight. Meanwhile, the other parties continued to work on ratifying the Kyoto Protocol.

Kyoto Protocol Approved

By late 2004, most of the world's countries had formally ratified the Kyoto Protocol. Important exceptions included the United States and Australia. Shortly after ratification by Russia, the protocol entered into force on February 16, 2005, which means it became legally binding on countries that approved it. This means that each country that ratified the Kyoto Protocol has until 2012 to reduce its emissions to the targeted amount. Governments can use a variety of flexible methods to reach their country-specific goals, such as carbon trading and investing in clean technologies.

The Road Ahead

Experts agree that while the Kyoto Protocol is a step in the right direction, it is only a beginning. Governments will need to take additional measures to address such a massive problem. The Fourth Assessment of the Intergovernmental Panel on Climate Change (IPCC), released in late 2007, removed any remaining doubt about human-induced climate change and provided a powerful mandate to take further action. Before the first Kyoto Protocol commitment period ends in 2012, governments must get a new agreement in place—one that can reduce emissions on a scale that matches the size of the problem.

To learn about the most recent developments regarding the Kyoto Protocol, as well as the United States government's actions regarding climate change, check out the web sites listed in the "Become Kyoto Savvy" section of Tip 50. In addition, the following sites will give you a more in-depth understanding of the Kyoto Protocol:

http://unfccc.int/kyoto_protocol/items/2830.php—Here is a good introduction to the Kyoto Protocol.

www.cckn.net/compendium/int_backgrounder.asp—This is an excellent overview of all the Conferences of the Parties. Explore the rest of this web site for in-depth information on the details of the Kyoto Protocol.

http://www.weathervane.rff.org/glossary.cfm—Check out this exhaustive glossary for all those confusing Kyoto words and acronyms.

Last, for more in-depth information on climate change policy, both present and future, check out this book by Niklas Höhne: *What Is Next After the Kyoto Protocol: Assessment of Options for International Climate Policy, Post 2012.*

Appendix C: Our Calculations, Units of Measure, and Internet Sites

1. A Word About Our Calculations

National Averages

All the calculations in this book were based on the most recent national averages available at press time. This means two things. First of all, because *recent* is a relative word and statistics become outdated with every passing second, your money and carbon dioxide savings could be higher or lower than our calculations predict. Second, because you probably don't live in a precisely 1,600-square-foot house, drive your car exactly 12,000 miles per year, or get charged by your utility company exactly 8.16 cents per kilowatt-hour of electricity, your savings will almost definitely be higher or lower than the savings we predict for each tip. Your money and energy savings will ultimately depend on many factors, including your climate, the size and age of your house, the number and lifestyles of the people living in your house, and the efficiency of your appliances and car.

Rate of Return?

As you may have noticed, we do not calculate the rate of return for our tips that require an initial investment. Instead, we simply calculate how much a particular tip will save the average household in 1 year and then multiply that by the number of years the tip will last. For example, in Tip 23, caulking and weather stripping save the average household $56.10 a year and last for 10 years, for a *total* savings of $561. From there, we subtract the initial cost of the product to find the net savings. In this example, the caulking and weather stripping cost $62, so $561 – $62 = $499 net savings. Finally, we divide the net savings by the number of years the tip will last. Since the caulking and weatherstripping will last for 10 years, we divide $499 by 10 years to get an *annual net* savings of $49.90 a year, which we round up to $50 a year.

2. A Word About Our Units of Measure

Here are some definitions to help you better understand the units of measure we use throughout this book.

Watt (W): Watts describe how much electricity is used during any given moment. A 100-watt lightbulb uses 100 watts of electricity at any given moment, whereas a 75-watt bulb uses 75 watts at any given moment.

Kilowatt (kW): 1 kilowatt = 1,000 watts.

Megawatt (MW): 1 megawatt = 1,000 kilowatts (or 1,000,000 watts). Megawatts are typically used to measure the electricity-generating capacity of power plants. The average U.S. power plant has the capacity to produce 213 megawatts of electricity at any given moment.

Watt-hours (Wh): A watt-hour is the measure of how much electricity is used (watts) over an hour. For example, a 100-watt light bulb, which uses 100 watts of electricity at any given moment, will use 100 watt-hours of electricity in 1 hour, 200 watt-hours of electricity in 2 hours, or 150 watt-hours in 1.5 hours.

Kilowatt-hours (kWh): 1 kilowatt-hour = 1,000 watt-hours. In other words, if you have an appliance that uses 1,000 watts of electricity at any given moment and plug it in for one hour, you've used 1 kilowatt-hour. Kilowatt-hours are used to measure the amount of electricity that households use each month, and they are what you see on your electricity bill. To find out how much your utility charges you per kilowatt-hour, simply divide your monthly bill by the total number of kilowatt-hours you used that month. The national average retail price of electricity in 2000 (which is what we used for this book's calculations) was 8.16 cents per kilowatt-hour.

Lumens: Put simply, lumens are the measure of light output. The lumen rating on a lightbulb indicates how much light the bulb will give off, while the watt rating indicates how much electricity it will use when it's on. The higher the lumens, the brighter the bulb will be.

British thermal unit (Btu): A British thermal unit, or Btu, is a measure of energy. More specifically, it is the amount of heat (energy) needed to raise the temperature of one pound of water 1°F. Kilowatt-hours, therms, joules, and many other units of energy can all be converted into Btus.

Therms: 1 therm = 1,000 Btus. Natural gas is typically measured in therms. Throughout this book, we use the accepted standard of one therm of natural gas burned emits 12 pounds of carbon dioxide into the atmosphere.

Miles per gallon (mpg): Miles per gallon, also known as a vehicle's fuel efficiency, is the number of miles your car can drive using 1 gallon of gasoline. The typical U.S. passenger car can drive 23 miles on 1 gallon of gas, so it gets 23 miles per gallon. The typical SUV gets 16 miles per gallon.

Pound of carbon dioxide: Gases are often measured in pounds (a measure of mass), although they can be measured in a variety of ways. To calculate carbon dioxide emissions, scientists first weigh the amount of fossil fuel being burned, then determine how much of that weight is carbon, and finally multiply that number by 3.66 to determine the weight in carbon dioxide. Alternatively, they can use a piece of equipment called an infrared gas analyzer to measure the concentration of carbon dioxide in the air, after which they can convert the carbon dioxide concentration into pounds. Throughout this book, we use the accepted standard of 1 kilowatt-hour of electricity used = 1.64 pounds of carbon dioxide emitted at the power plant.

Ton of carbon dioxide: 1 ton of carbon dioxide = 2,000 pounds of carbon dioxide.

3. A Word About Our Internet Sites

Many of the Internet sites we recommend are in pdf format, which means they require the use of software called Acrobat Reader. You can download a recent version of Acrobat Reader for *free* from this web site: www.adobe.com/products/acrobat/readstep2.html. Also, if you don't want to type in all the long Internet addresses referenced in this book, you can easily access them through the book's web site: www.preventglobalwarming.net.

Disclaimer 1: Some of the Internet sites we recommend may not be the prettiest or most professional-looking you've ever seen, but they are—in our opinion—the best of what's out there in terms of content.

Disclaimer 2: We have attempted to contact the managers of all the Internet sites we recommend in this book, asking them not to change the URL's (Internet addresses) of their pages. If they absolutely must change their URL's, we asked if they would at least create a "redirection" link so that our readers wouldn't encounter broken links. This is all we can do—the Internet is constantly changing and being updated. If you cannot find some of the sites we recommend, we sincerely apologize.

Appendix D: Tips Especially for Renters

1. Take It with You

The vast majority of tips in this book *do* apply to renters, although we realize that there may be some things that are out of your control, such as the temperature of your hot-water heater or the amount of insulation in your roof. There are two main strategies to follow if you want to conserve energy and save money as a renter. The first is to buy things you can take with you to your next apartment or house. These include

- Indoor drying rack for clothes (Tip 3)
- Programmable thermostat (Tip 19)
- Window air conditioners, ceiling fans, window fans (Tips 20 and 21)
- Toilet dams and/or flappers (Tip 7)
- Compact fluorescent lightbulbs (Tip 1)
- Low-flow showerheads, faucet aerators (Tip 6)
- Filtered water bottle, faucet water filter (Tip 8)
- Curtains, blinds, shutters, et cetera (Tip 22)
- New refrigerator, washing machine, dishwasher, or computer (Tips 14 and 15)
- Storm windows, sun-control screens (Tip 27)

2. Focus on What You Pay For

The second thing you should do if you're a renter is determine which utilities (electricity, hot water, heat, and so on) you pay for directly and concentrate on the tips that apply to those utilities. Once you've done everything you can on your own, ask your landlord or superintendent (or gain support from your fellow apartment dwellers and then ask your super) to do things such as fix leaky heating ducts, increase insulation, or add an insulating jacket to the building's water heater. Also, demand that he or she fix specific problems in your apartment, such as leaky faucets or radiators that give off so much heat you're forced to open a window.

For any tip in this book over which you don't have any control, ask your landlord or superintendent to do it. Even though it won't earn him or her any extra money (the utilities get your money, not the landlord), it won't *hurt* him or her either. Another option would be for you and your fellow tenants to pool your money to make the initial investments, after which you would all reap the savings from lower utility bills.